TECHNICAL
REPORT

Information Systems Technician Rating Stakeholders

Implications for Effective Performance

Margaret C. Harrell, Harry J. Thie,
Roland J. Yardley, Maria C. Lytell

Prepared for the United States Navy

RAND NATIONAL DEFENSE RESEARCH INSTITUTE

The research described in this report was prepared for the United States Navy. The research was conducted within the RAND National Defense Research Institute, a federally funded research and development center sponsored by the Office of the Secretary of Defense, the Joint Staff, the Unified Combatant Commands, the Navy, the Marine Corps, the defense agencies, and the defense Intelligence Community under Contract W74V8H-06-C-0002.

Library of Congress Control Number: 2011942594

ISBN: 978-0-8330-5941-3

Published 2011 by the RAND Corporation
1776 Main Street, P.O. Box 2138, Santa Monica, CA 90407-2138
1200 South Hayes Street, Arlington, VA 22202-5050
4570 Fifth Avenue, Suite 600, Pittsburgh, PA 15213-2665
RAND URL: http://www.rand.org/
To order RAND documents or to obtain additional information, contact
Distribution Services: Telephone: (310) 451-7002;
Fax: (310) 451-6915; Email: order@rand.org

Preface

In a prior study (Thie et al., 2009), the U.S. Navy Program Executive Officer of Command, Control, Communications, Computers and Intelligence (PEO C4I) asked the RAND Corporation to assess broadly the manpower, personnel, and training implications associated with the introduction of the Consolidated Afloat Networks and Enterprise Services (CANES) program aboard ships. The CANES program is designed to consolidate and improve the networks on tactical platforms, largely through a common computing environment, and was presumed to have implications especially for information technology personnel. That prior study concluded that the Navy lacks a single "czar" who can integrate technological change, ship structure, manpower/manning/training processes, and operational practice to produce change. As a result, gains in efficiencies and effectiveness from implementing the CANES program will be difficult to realize fully without close collaboration and alignment of interests among the stakeholders.

PEO C4I asked RAND to undertake a further study to help better understand the relationships among stakeholders involved in manpower, personnel, and training processes for managing the information systems technician (IT) rating and the effects of these different stakeholders upon efficiency and effectiveness measures for the community in general and the CANES program specifically. This report addresses those issues. The report assumes some knowledge of military manpower, personnel, and training processes and should be of interest to Navy and Department of Defense officials involved in manpower, personnel, and training.

This research was conducted within the Forces and Resources Policy Center of the RAND National Defense Research Institute, a federally funded research and development center sponsored by the Office of the Secretary of Defense, the Joint Staff, the Unified Combatant Commands, the Navy, the Marine Corps, the defense agencies, and the defense Intelligence Community.

For more information on the RAND Forces and Resources Policy Center, see http://www.rand.org/nsrd/ndri/centers/frp.html or contact the director (contact information is provided on the web page).

Contents

Figures

Tables

Summary

Introduction

The U.S. Navy is organized by five warfare enterprises and with various providers that contribute services, materiel, and human resources to achieve overall readiness. This system is characterized by complex networks of organizations with disparate goals and priorities. For example, the relative emphases on efficiency and effectiveness vary throughout this enterprise system.

The CANES initiative is designed to consolidate and improve the networks on tactical platforms with a common computing environment. This consolidation was intended to result in greater training and operating effectiveness, specifically for information technology enlisted personnel. The greater effectiveness could increase performance or be translated into efficiency savings. However, preceding research (Thie et al., 2009) indicated that the complex pattern of stakeholders involved in the manpower, training, and personnel processes for the IT rating confounded the pursuit of increased effectiveness.

This report describes the relationship among stakeholders involved in the manpower, personnel, and training processes for information technology personnel. For comparison purposes, this report also describes the stakeholder relationships for two other Navy ratings (hospital corpsman and Aegis firecontrolman), as well as an IT rating counterpart in the Air Force, cyber operations.

This effort is intended to contribute to maximizing the performance outcomes from technology insertion, create greater awareness of and support for effectiveness as an objective, highlight the conflicting stakeholder patterns that shift decisions away from effectiveness; and recognize decision tradeoffs. Further, the intent of this report is to provide insights for IT stakeholders regarding the organizations with which they interact, and to facilitate better interaction among those stakeholders.

Stakeholders and Ratings

This report adapts stakeholder theory to the issue of IT rating "ownership" in the Navy and, thus, interprets stakeholders as individuals or groups that can affect, or that are affected by, an organization's actions in pursuing its goals. We specifically focus on the stakeholders that affect (or that are affected by) the manpower, personnel, and training (MPT) decisions for the ratings considered. Because the landscape of stakeholders can be extensive, we focus primarily on a subset of those IT stakeholders.

There are many stakeholders that influence MPT decisions for the IT rating. Overall, the IT rating lacks "ownership" by any single organization or individual. This lack of ownership is further confounded by the distribution of IT personnel in the Navy. Typically IT personnel are assigned in relatively small numbers to many different units and types of units, to include units represented within four different Type Commanders (TYCOMs) (Naval Air Forces, Naval Surface Forces, Submarine Forces, Navy Cyber Forces). Thus, any performance outcome for IT personnel is diluted among many different organizations, and there is no concerted voice regarding IT rating management or acting on behalf of IT personnel. The other ratings we reviewed provide distinct stakeholder pattern comparisons to that of the IT rating.

An Aegis firecontrolman (FC) follows similar MPT processes as does an IT, but the IT rating stakeholder patterns have a couple of key differences. First, the resource sponsor for new equipment training for the FC rating is N86, a well-established office. In comparison, Office of the Deputy Chief of Naval Operations for Information Dominance (N2/N6), a newly established office, is the resource sponsor for new equipment training for the IT rating. Although the FC rating also lacks clear ownership by a single stakeholder, there is a key difference between the IT and the FC rating. FC performance affects only a single TYCOM user: Naval Surface Forces. As a result, and because an FC operates a combat warfare system, FC performance is emphasized to a greater degree than is the case for the IT rating, and recent revisions to FC training to accompany the new Aegis system have received the necessary funding and support.

Hospital corpsman (HM) has a very different stakeholder pattern from that of IT, since the HM rating is managed with the "czar" leadership model, with the Navy Bureau of Medicine and Surgery (BUMED) actively involved in all MPT decisions. BUMED personnel are involved in accessing, training, and assigning each HM. As both the primary user and also the clear owner of the HM rating, BUMED is capable of and enabled to set efficiency and effectiveness priorities, such as decisions to man an organization with fewer, better-trained, individuals. This stakeholder pattern is especially of interest to the IT rating because the first indication of the Information Dominance Corps suggested that the IT rating might be managed within the Information Dominance Corps with a "czarship" model similar to that used to manage HM.

The Air Force management of cyber operations personnel follows a different pattern. As for the HM rating, there is clear ownership of the occupation; the Air Force's Chief Information Officer is the functional authority for this occupation and is involved in MPT decisions. Although the Air Force cyber operations community is not managed by a czar, the major command (MAJCOM) user is involved in training decisions and has the opportunity to fund additional individual training to benefit the MAJCOM. Such effectiveness and efficiency decisions (spending user resources to improve individual and systemwide performance) are not possible in the MPT processes for the IT or the FC rating because of defined stakeholder responsibilities.

Our review of three Navy ratings and one Air Force Specialty Code led us to the observations in Table S.1. We recognize that the observations are based on a limited sample and make no claims that they are generalizable across all occupations. Moreover, in terms of favorable program (e.g., MPT) outcomes based on practices and policies for these occupations, we rely on interviews and our own assessments. There are differences in policy, practice, and outcomes across the four occupations, particularly within the Navy and between the Navy and the Air Force.

Even this limited sample illustrates significant differences within the Navy and between the Navy and the Air Force. In our scoring system, indicated in Table S.1, a check mark represents at least partial agreement with the observation. Some of these observations are for condi-

Table S.1
Observations Regarding the Four Communities

Observation	IT	FC	HM	Air Force Cyber Systems Operations
With Respect to Power				
Focus on efficiency and effectiveness, with ability to make tradeoffs between the two			√	√
The ultimate decisionmakers: users, rather than program and budget providers			√	√
With Respect to Legitimacy				
Training emphasized, with an identified functional lead, regular reviews, and hard metrics			√	√
A clear "ownership" of the specialty, with ultimate authority and central decisionmaking for policy and directives related to the community			√	√
Considerable influence in training and budgeting decisions by those with expertise in the career field			√	√
Fewer stakeholders involved; easy to gain consensus			√	√
A smooth transition of new equipment training into steady-state training			√	√
With Respect to Urgency				
Specialty concentrated: assigned to units in large numbers		√	√	Not observed
Central involvement in decisionmaking for the occupation by the units that benefit from performance improvements or suffer from decreases in performance			√	√
Performance prioritized		√	√	√

tions outside the control of the manpower, personnel, and training community (e.g., where and how used). However, most observations should be within the control of these communities.

Effectiveness and Efficiency Objectives

Goals or objectives of organizations can be considered within an effectiveness and efficiency framework. *Efficiency* is a primary focus on inputs, use of resources, and costs. *Effectiveness* features a primary focus on outputs, products or services, and growth. Organizational strategies can be located within these two dimensions (Burton, DeSanctis, and Obel, 2006).

An organization must make decisions along two dimensions of goals: effectiveness and efficiency. Goals may be in conflict if the organizational structure that fixes stakeholder decisionmaking leads some parts of the organization to focus on effectiveness, without considering efficiency, and other parts to focus on efficiency, without considering effectiveness. There is a potential tension between effectiveness and efficiency that is not an inherent conflict but, instead, one created by organizational structure and decisionmaking patterns. Our assessment in Chapter Four shows that all of the organizational stakeholders that we assessed have an efficiency strategy but that not all of them have a positive effectiveness strategy.

Recommendations

Navy leadership should be cognizant of the goal orientation and strategy of organizations in order to assess whether those goal orientations are the most appropriate for organizations. Further, awareness of differing goal orientations can facilitate interactions by bringing explicit awareness of differing stakeholder strategies. Our recommendations are specific to the IT community but might also apply to other Navy communities.

The IT community should have a single identified stakeholder with responsibility for performance and MPT oversight. Ideally, this stakeholder would have sufficient legitimacy and power (i.e., money) to influence change. Since the IT community has many different users, this predominant stakeholder should interact with all users as part of an institutional process for gathering feedback from users and making improvements. Organizationally, who should that be and how could it be implemented? For the HM community, the BUMED organization has the legitimacy of the Office of the Chief of Naval Operations instruction authority; the "power of the purse" (resourcing of training); and urgency, as the user of the outcome of training. This model has appeal, and early documentation for the creation of an Information Dominance Corps cited such a model, but it has not eventuated. The FC model appears currently workable for that community because of a strong system command, resource sponsor, and a TYCOM with stable, consistent relationships. The IT stakeholders—whether enterprise, TYCOM, or Office of the Chief of Naval Operations staff—have changed structural relationships considerably in recent years. Absent radical change either to an HM model or change of a type we did not consider,[1] an Air Force model might offer the best path to effectiveness and efficiency. The N2/N6 would be the dominant stakeholder in this model and would be responsible for development and training of IT personnel. Similar to the Air Force model, this authority would be supported by functional authorities within N2/N6 and by a career field manager in an operational unit with authorities designated in Navy instructions. The N2/N6 would have ultimate responsibility for effectiveness of the community. This arrangement would need to be supported by more-workable processes than currently exist.

For example, there should be regularly scheduled, periodic reviews of IT sustainment and new equipment training. For this community, a periodicity of 18–24 months (rather than three years) may be appropriate because of the rate of expected change in technology. As with the Air Force, these reviews should be under the auspices of the functional lead and chaired by the functional career manager with support provided by NETC and the schoolhouse. These reviews should include users, and have the oversight of the N2/N6, the predominant stakeholder for decisionmaking and resourcing. Such reviews and resulting periodic change could move the IT MPT communities toward effectiveness as an outcome.

---

[1] For example, N2/N6 assigned training resource provider and TYCOM responsibilities with SPAWAR as training provider. These could have merit to consider. In particular, because the IT community appears to have more technology change and thus more new equipment training than many other communities, it may increase effectiveness and efficiency to have one organization responsible for resourcing and executing both sustainment and new equipment training.

At the type commander level, IT users should have the opportunity and capability to make efficiency and effectiveness decisions. For example, the opportunity to have fewer, better-trained personnel should be one such decision regularly considered. Because of the number and diffused nature of the user stakeholders, the functional authority (N2/N6) needs to take the prevalent role in these manpower reviews supported by the N12 manpower community.

While these recommendations are consistent with good management practices, we recognize that it will be difficult to improve the effectiveness of IT personnel for several reasons. First, because they are generally diffused throughout the Navy in small numbers, there may not be users who will see significant organizational performance increases because of increased IT performance. Second, IT personnel do not operate or maintain a Navy combat system. Unlike the Air Force, which has recognized and publicly stated that information technology is vital to their mission performance, the Navy has not prioritized IT personnel performance.[2] Given the institutional emphasis on efficiency foremost, additional investments in IT performance are currently unlikely without significant changes in organizational structure and processes and in stakeholder responsibilities.

---

2 The Chief of Staff of the Air Force has stated that "cyber operations reinforce and enable everything we do, from administrative functions to combat operations, and we must treat our computers and networks similarly to our aircraft, satellites, and missiles" (Lieutenant General William Lord's introduction to *Cyberspace—Shaping the New Frontier*, 2010). The Navy considers its networks to be in support of Navy weapons systems.

Acknowledgments

We would first like to thank and recognize the many Navy and Air Force personnel who agreed to interviews with us during the course of this research. This study would not have been possible without their generosity of time and candor regarding their experiences and insights. Our immediate research sponsor, Sean Zion, patiently supported our efforts as we conducted this and the prior research. Within RAND, we benefited from helpful reviews from Tom Manacapilli, John Schank, and Michael Hansen.

Abbreviations

A1	Air Force Deputy Chief of Staff for Manpower and Personnel
AETC	Air Education and Training Command
AF	U.S. Air Force
AFRS	Air Force Recruiting Service
AFSC	Air Force specialty code
ASVAB	Armed Services Vocational Aptitude Battery
ATRC	Aegis Training and Readiness Center
BUMED	Navy Bureau of Medicine and Surgery
C4I	Command, Control, Communications, Computers, and Intelligence
C4ISR	Command, Control, Communications, Computers, Intelligence, Surveillance, and Reconnaissance
CANES	Consolidated Afloat Networks and Enterprise Services
CASREPs	casualty reports
CFETP	Career Field Education and Training Plan
CFM	career field manager
CG	cruiser
CID	Center for Information Dominance
CIO	Chief Information Officer
CMC	Commandant of the Marine Corps
CNP	Chief of Naval Personnel
CSCS	Center for Surface Combat Systems
CVN	carrier
DASAF	Deputy Assistant Secretary Air Force

DDG	destroyer
DoD	Department of Defense
FA	functional authority
FC	firecontrolman
FEA	front-end analysis
FLTFORCOM	Fleet Forces Command
FM	functional manager
HM	hospital corpsman
HR	human resources
ISEA	in-service engineering agent
IT	information systems technician
IWS	Integrated Warfare Systems
KSA	knowledge, skills, and abilities
MAJCOM	major command
MFM	MAJCOM functional manager
MPT	manpower, personnel, and training
MPT&E	Manpower, Personnel, Training, and Education
N1	Office of the Deputy Chief of Naval Operations for Manpower, Personnel, Education and Training
N12	Total Force Requirements Division of N1
N15	Education and Training Division of N1
N2/N6	Office of the Deputy Chief of Naval Operations for Information Dominance
N8	Office of the Deputy Chief of Naval Operations for Integration of Capabilities and Resources
N86	Surface Warfare Division of N8
NAVMAC	Navy Manpower Analysis Center
NAVNETWARCOM	Naval Network Warfare Command
NEC	Navy enlisted classification
NETC	Naval Education and Training Command
NIDE	Naval Information Dominance Enterprise

NNFE	NETWAR/FORCENet Enterprise
NPC	Navy Personnel Command
NPDC	Naval Personnel Development Command
NRC	Naval Recruiting Command
NTSP	Navy Training System Plan
OD/OC	organizational development and organizational change
OPNAV	Office of the Chief of Naval Operations
OSD	Office of the Secretary of Defense
PEO	Program Executive Officer
PM	program manager
PMW	Program Manager Warfare
POM	program objective memorandum
PPBE	planning, programming, budgeting, and execution
RFT	ready for training
ROC	required operational capability
SECNAV	Secretary of the Navy
SHRM	strategic human resource management
SME	subject matter expert
SPAWAR	Space and Naval Warfare Systems Command
SPECWARCOM	U.S. Navy Special Warfare Command
SSA	software support activity
SYSCOM	systems commanders
TA	training agency
TEE	training effectiveness evaluation
TM	training manager
TPM	training pipeline manager
TSA	training support agency
TYCOM	Type Commander
U&TW	Utilization and Training Workshop
USFFC	U.S. Fleet Forces Command

CHAPTER ONE

Introduction

Background

The U.S. Navy uses an enterprise construct to manage its business.[1] One of its stated goals is to improve output (effectiveness) over cost (efficiency). The Naval Information Dominance Enterprise (NIDE) (formerly NETWAR/FORCENet Enterprise [NNFE]) is one of five warfare enterprises that, at the time of our research, comprised the Fleet Readiness Enterprise.[2] NIDE/NNFE has been in existence for about four years and consists of commands involved in the business of Command, Control, Communications, Computers, and Intelligence (C4I) and information operations, such as the Space and Naval Warfare Systems Command (SPAWAR) and the Program Executive Officer for Command, Control, Communications, Computers, and Intelligence (PEO C4I). In the enterprise construct, providers contribute services, materiel, and human resources to the enterprises to achieve readiness. In particular, Manpower, Personnel, Training and Education (MPT&E) is a provider. SPAWAR and PEO C4I are also providers to other enterprises, such as the Surface Warfare Enterprise. Even at this level, the relationships are complex for decisionmaking and delivery of output effectiveness.

Moreover, the providers are complex networks of organizations and have relationships themselves. For example, MPT&E includes policy organizations (the Secretary of the Navy [SECNAV] and various Office of the Chief of Naval Operations [OPNAV] staffs) and implementing organizations (e.g., the Naval Education and Training Command [NETC] and the Navy Manpower Analysis Center [NAVMAC], and the Navy Personnel Command [NPC]). Even within these staffs and organizations, disparate and/or conflicting goals and priorities exist. For example, cost and efficiency may dominate the consideration of one stakeholder while effectiveness may be the metric for another.

The need to integrate disparate goals is evident in the development of the Consolidated Afloat Networks and Enterprise Services (CANES) program. The various C4I and warfare sys-

[1] "In March 2008, the Navy Enterprise construct was described by Navy Enterprise Chief Operating Officer RADM David Buss as 'activities, governance, and behaviors that will drive additional efficiencies in how the Navy delivers current readiness and future capability, as well as provide a foundation for making better, more informed mission, capability, resource allocation, and risk decisions.' Today, the purpose of the Navy Enterprise construct, as stated by the same source, is to achieve additional efficiencies so that current and future readiness can be met with limited budgets. More specifically, the Navy Enterprise seeks to gain an improved return on investments by 'improving output over cost' and 'improving resource allocation effectiveness'" (Riposo et al., 2009, p. 1).

[2] The Fleet Integration Executive Panel, chaired by both the Commander U.S. Pacific Fleet and the Commander U.S. Fleet Forces Command, has replaced the Fleet Readiness Enterprise. The NNFE and NIDE are slightly different from one another because of the recomposition of the Office of the Deputy Chief of Naval Operations for Information Dominance (N2/N6) portfolio.

tems on naval ships and at shore installations are currently developed, fielded, and supported largely independently of one another. The result has been numerous stovepiped networks, each with unique hardware and software systems administered and maintained by enlisted personnel in the information systems technician (IT) rating. The CANES initiative is designed to consolidate and improve the networks on tactical platforms, largely through a common computing environment.

This consolidation and improvement should result in greater training and operating efficiency and effectiveness for IT personnel, which may be parlayed into manpower savings (cost reduction) or increased personnel effectiveness or a balance of the two, thus improving output over cost. However, preceding RAND research indicated that multiple stakeholders with disparate goals are involved in the accessing (entry into service), training, management, utilization, and career progression of IT personnel. This research will clarify the different stakeholders involved in the manpower, personnel, and training (MPT) processes for IT personnel; will delineate their goal orientation; and will seek opportunities to identify, and even facilitate, increased IT personnel effectiveness.

Research Questions

The RAND Corporation was asked to conduct research to understand the relationships among stakeholders involved in manpower, personnel, training, and education processes for managing the IT rating and the implications of these different stakeholders for cost and effectiveness outputs of the community in general and the CANES program specifically.

We conducted this research by first reviewing the literature with respect to stakeholder theory, organizational design, and organizational decisionmaking. Following that, we used past research and interviews to identify stakeholders and restrictions to MPT improvements. Our intent was to identify those organizations and offices (across enterprises, within the NIDE/NNFE enterprise, and within the MPT&E providers) that influence, or are affected by, the manpower, career development, training, and assignment of IT personnel. These interviews also permitted RAND to portray the priorities and perspectives of the stakeholders. Analysis of the stakeholder perspectives permitted RAND to identify the extent to which this stakeholder pattern restricts, or contributes to, improvements to IT personnel effectiveness and management.

We also asked about comparable Navy ratings and conducted interviews in order to identify stakeholder patterns across the comparison ratings. These interviews and analyses allowed us to assess whether the restrictions apparent from the stakeholder pattern for IT personnel are a unique conundrum or whether there are similar issues across other Navy ratings. In a similar fashion, we conducted interviews with personnel who train and manage comparable technical personnel in the U.S. Air Force (AF) in order to identify useful lessons.

Last, our conclusions and recommendations focus on the conflicts and resulting restrictions apparent among Navy IT stakeholders, the extent to which these relationships are unique to the Navy IT community, and the tradeoffs and compromises necessary to improve the training, management, and resulting performance of IT personnel.

Organization of the Report

Chapter Two of this report provides the theoretical basis—including a discussion of stakeholder theory, organizational development and change, and organizational fit. Chapter Three assesses IT and selected comparison ratings. Chapter Four discusses the IT community stakeholders and their efficiency and effectiveness strategies. The final chapter provides conclusions and recommendations. This report also includes an appendix that provides more detail on Navy training processes and stakeholders and another appendix regarding the methodology used to calculate an effectiveness and efficiency goal orientation for the IT stakeholders.

CHAPTER TWO

Theory from the Literature

This chapter reviews the literature concerning stakeholder theory, human resource management strategy, management structures for training, and organizational design. This review introduces concepts that will be used throughout the report.

Stakeholder Theory Literature

In the strategic management literature, there is a research stream defined by what is called "stakeholder theory." Its emergence as a research stream is credited to Freeman's (1984) book titled *Strategic Management: A Stakeholder Approach.* The initial focus of this theory was to provide organizational managers with a way to take into account all relevant stakeholders, not just shareholders, when making strategic decisions. Over the years, stakeholder theory has grown into a set of theories that differ in terms of rationales for why organizational managers should care about stakeholders. The proliferation of theories has thus led to a proliferation of definitions of *stakeholders.* One of the broadest, and most widely used, definitions of stakeholders comes from Freeman, who defined stakeholders as parties who can have an impact on, or who are impacted by, an organization as its leaders take actions in pursuit of organizational goals. Narrower definitions of stakeholders usually place qualifiers on the characteristics of the parties, most notably the qualifier that stakeholders are parties with legitimate claims on the organization (e.g., those who have invested resources in the organization).

Identifying Stakeholders

A somewhat separate, but related, issue to the definition of stakeholders is the problem of identifying stakeholders. Freeman (1984) classified stakeholders into four groups by crossing two factors: location (internal versus external) to the organization and nature of interests in the organization (cooperative versus competitive). Based on his typology, Freeman then advocated four generic strategies for dealing with each type of stakeholder. In their recent review of the stakeholder theory literature, Laplume, Sonpar, and Litz (2008) pointed out that other theorists have proposed additional classifications over the years, with some authors arguing that current classifications of stakeholders fail to explain the heterogeneity of different stakeholder groups.

Perhaps a more feasible approach to identifying stakeholders has been to focus on stakeholders that managers *actually* pay attention to, not which stakeholders that managers *should* pay attention to. One example of this approach has been the model of stakeholder salience proposed by Mitchell, Agle, and Wood (1997). Mitchell and colleagues argue that stakeholders will be more salient to managers the more they are believed to possess the following characteristics: power, legitimacy, and urgency. Based on all combinations of these three perceived characteristics, the authors proposed a typology of seven stakeholder entities. This model has received some empirical support with data from private, for-profit organizations (e.g., Agle, Mitchell, and Sonnenfeld, 1999).

In general, the literature on stakeholder theory does not offer a single, best approach to defining and identifying stakeholders. Part of the reason for this is that organizations experience different environmental constraints, thus making some stakeholders more important in some contexts than in others (Carroll and Hannan, 2000). Another reason is that some of the models that have been tested on actual organizations used only one type of organization, typically large, for-profit organizations (Laplume, Sonpar, and Litz, 2008). These limitations notwithstanding, the stakeholder-theory literature provides useful concepts for defining stakeholders and has at least one empirically tested framework—Mitchell, Agle, and Wood's (1997) model of stakeholder salience—for identifying relevant stakeholders.

Implications for the Current Study

We began our study by developing a definition of stakeholders based on the specific interest in IT community "ownership" in the Navy. Using Freeman's (1984) definition as a starting point, one could define a Navy IT community stakeholder as a party (i.e., an individual or a group) that has an impact on MPT decisions for IT personnel or that is impacted by such decisions. Based on this definition, a number of parties who make IT MPT decisions have been identified (e.g., the Office of the Deputy Chief of Naval Operations for Manpower, Personnel, Education and Training [N1]; N2/N6; and NAVMAC). Relative to parties that have an impact on IT MPT decisions, parties that are impacted by IT personnel decisions in the Navy have been under-identified. Because IT personnel are used broadly across Navy organizations and units but for the most part in small numbers in each unit, the user community has not been a dominant stakeholder. Said another way, the stakeholders dealing with inputs and the efficiency of them have been more prominent than the stakeholders dealing with outputs or effectiveness. The main group that has not been explicitly identified as a stakeholder group consists of the IT personnel themselves. IT personnel may provide valuable insights about the impact of MPT decisions on the work of IT personnel on Navy ships.

For this research, we define *legitimacy* as the result of designations in OPNAV and other Navy Instructions, *power* as that of the purse (or resource sponsor in Navy terms), and *urgency* as the fleet or customer need. In Chapter Three, we use the stakeholder definitions and observations about power, legitimacy, and urgency to organize our assessment of communities.

Strategic Human Resource Management and Management Structure

A large research stream in management is the strategic human resource management (SHRM) literature, which focuses on identifying and describing how an organization's human resources (HR) strategies relate to the organization's performance. HR principles, policies, practices, and products concern many personnel-related issues in organizations, such as staffing, compensation, training, recruiting, and performance management. Like the stakeholder-theory literature, the SHRM literature has a lot more theory than systematic empirical research. However, there are some things that can be learned from the SHRM literature and applied to the context of this study.

Based on a review of the SHRM literature by Legnick-Hall et al. (2009), a number of studies have shown that HR strategies contribute positively to an organization's performance (particularly financial performance). However, there is a lot less empirical evidence showing *how* HR strategies relate to an organization's performance. Legnick-Hall and colleagues suggested that the multilevel theories of SHRM that have been developed over the past decade seem the most promising for figuring out what is in the "black box" between HR strategies and organizational performance. An example of one of these multilevel theories was proposed by Bowen and Ostroff (2004), who argued that HR systems that create a "strong situation" will result in organizational members interpreting HR policies and practices similarly. In turn, consensus thinking about the organization's HR policies and practices will result in consistent and positive individual-level behaviors that will predict positive organization-level performance. More specifically, Bowen and Ostroff made the case for organizational climate, or shared perceptions of contingencies between behaviors and outcomes in the organization, as the means through which employee behaviors translate into organization-level performance. These authors thus argue that HR systems that are high in distinctiveness, consistency, and consensus "should enhance clarity of interpretation in the setting . . . whereby individuals yield to the message and understand the appropriate ways of behaving" (Bowen and Ostroff, 2004, p. 213).

Although not specifically a part of the SHRM literature, a RAND Corporation report by Manacapilli et al. (2004) addressed the issue of organizational management structure and information flow as it relates to training (an HR function) in the AF. The study was carried out in order to "help improve the responsiveness of the Air Force's technical training system to changing force and support requirements" (p. 13). Manacapilli and his team at RAND used a "best practices" approach by examining how other organizations, both military and civilian, manage their training programs. Based on their reviews of other training programs and the theoretical literature on organizational management structure, the RAND researchers proposed an exemplar of a training organization with different levels of management. The researchers then compared how the AF technical training structure fits within the exemplary training management structure and how training data should flow between management levels. The researchers then conducted a "deficiency analysis" by examining how four types of training data (i.e., cost, capacity, quantity, and quality data) flow between the AF training functions, which were grouped by the proposed management levels. This deficiency analysis revealed that certain types of training data were either not available or not accessible (i.e., understandable) to certain management levels that should have the data made available and accessible to them. Thus, by doing this analysis, Manacapilli and his team were able to point to specific areas where the AF can improve its training information flows and to recommend ways in which the AF can restructure its training management functions based on reviews and metrics. We use these concepts in our assessment in Chapter Three.

Both the SHRM literature and the RAND Corporation report on management structure of technical training in the AF point to some general "best practices" for HR systems. First, decisionmaking should be decentralized (Ericksen and Dyer, 2005) by creating a flatter organizational structure. However, decisionmaking should not be decentralized to the point where stovepiping becomes a major problem. Manacapilli and colleagues specifically pointed to the Navy's training management structure (as of 2002) as an example of "over-decentralization" (pp. 15–16). Second, financial and performance information should be shared across many levels in the organization (i.e., increased information flow) so decisionmaking can be based on complete information. Third, organizations should try to use selective hiring practices and provide "extensive training" to their employees (Ericksen and Dyer, 2005, p. 925). Selective hiring and extensive training are particularly important in "high reliability organizations," for which highly reliable performance is required because of the volatility of the external environment and the high costs of making mistakes (e.g., as in nuclear power plants, hospital emergency centers, and Naval aircraft carriers) (Ericksen and Dyer, 2005, p. 908). Finally, HR systems need to have clear lines of responsibility for decisionmaking. That is, employees need to know who has authority over what decisions and "where the buck stops." For example, Manacapilli and colleagues made the suggestion of having one senior leader be responsible for organizing training and making strategic decisions, as is done in the Army's formal training structure (the Training and Doctrine Command). The insights from Ericksen and Dyer (2005) and from Manacapilli et al. (2004) inform our assessment in Chapter Three.

Although some HR best practices exist, implementing them in organizations can be a challenge. The literature on organizational development and organizational change (OD/OC) points to some of the problems involved in trying to make large-scale changes in organizations.

Organizational Development and Organizational Change

The OD/OC literature is concerned with the theory and practice of planned interventions in organizations. The literature on OD/OC has generally fallen into two camps: those who study the process of change in organizations (usually academics) and those who study how to implement programs for change in organizations (usually practitioners). These two camps do not often meet, nor do they often agree. Also, the methods used to validate theories vary between the two camps, with the change process researchers preferring hypothesis-testing methods and the implementation researchers preferring case studies (Austin and Bartunek, 2003). Having two camps in the OD/OC field has resulted in a proliferation of large-scale organizational interventions not being adequately tested before implementation.

Another, even more problematic concern in the OD/OC literature is whether large-scale change is even possible in organizations. A group of researchers in organizational demography (a subfield of sociology) promotes a theory of structural inertia, such that an organization's routines for providing reliable and accountable performance create inertia, making it very difficult for the organization to change (e.g., Amburgey, Kelly, and Barnett, 1993; Hannan and Freeman, 1984). As a result, organizational change is a function of selective forces in the external environment, with organizations with old forms failing and organizations with new forms being established. The proposed reason why it is difficult to change an organization without it failing is because the *process* of change, not so much the *content* of change, creates disruptions to the normal flow of activities and patterns of communication and decisionmaking in organizations (Barnett and Carroll, 1995).

The proposed reason that change does not often occur because change, in and of itself, is disruptive aligns with the concept of "resistance to change." Proponents of resistance to change argue that people resist change for a variety of reasons, including feelings of uncertainty about the change, fears about one's job security, lack of trust in those making the changes, certain personality characteristics, et cetera. Authors of management texts who address resistance to change often provide managerial strategies for overcoming employees' resistance to change (e.g., allowing employees to participate in the decisionmaking process). Dent and Goldberg (1999) argue that the idea that employees are naturally prone to resisting change unfairly puts the burden on employees to change and does not account for other sources of resistance, namely management and organizational systems. These authors thus proposed ditching the term "resistance to change," given its current interpretation as an employee-related phenomenon.

Organizational Culture

Perhaps a more overarching construct that can explain why change in organizations can be difficult is organizational culture. *Organizational culture* can be simply defined as "organizationally embedded assumptions and values" (Bowen and Ostroff, 2004). These assumptions and values often originate with the founding of the organization, such that the founding members embed their mission with certain assumptions and values that become entrenched via symbols (e.g., company logo), stories (e.g., how the company was founded), and structures (e.g., hierarchy of management levels). When a proposed organizational change threatens one of these cultural assumptions or values, resistance from multiple parts of the organization can arise. As Barnett and Carroll (1995) argued, changes to the structural core of the organization prove most disruptive. Organizational culture is the basis of an organization's structural core (Ostroff, Kinicki, and Tamkins, 2003). Cultural inertia is as much an issue as structural inertia.

Implications for the Current Study

Even if the Navy wanted to implement changes to its HR (i.e., MPT) systems, could it do so successfully? The answer is "maybe." There are studies that show that organizational change interventions have worked in both public- and private-sector organizations (e.g., Robertson and Seneviratne, 1995). Yet, the question is, How did those change interventions work? Specifically, how can one manage to change part or all of an organization's culture? An answer might come from a study by Jansen (2004), who proposed and tested a model of "change-based momentum" in a unique setting, the U.S. Military Academy (West Point). Jansen's ten-month study revealed that changes to organizational culture can occur (at least after initial implementation), but only if there is enough initial change-based momentum to overcome stasis-based momentum (i.e., persistence to staying the same course—inertia) and if other major events do not detract from the change event. A main take-away message from Jansen's study is that senior leadership needs to be intimately involved in making change happen. Senior leaders need to introduce the proposed change(s) in a compelling way (i.e., with what Jansen describes as urgency, feasibility, and drama), and they have to continue to be involved throughout the process so that individuals do not become distracted by other events. In Chapter Three, we observe how the AF has changed its training and development practice, and our recommendations suggest change for the Navy.

Organizational Fit

Most organizational researchers assert that goal orientation is an important component of an organization's design. Burton, DeSanctis, and Obel (2006), in particular, conclude that goal orientation, strategy, and environment must all "fit." For goal orientation, the important question is whether the organization is more focused on product (effectiveness) or process (efficiency). Knowing the goal orientation of an organization is a key input to defining the strategy of an organization. Strategies are either explorative or exploitative. For example, those companies that take initiative are explorers, whereas those that follow rules are exploiters. Other factors that define a strategy include whether the organization places greater value on expertise or collaboration. Likewise, are decisions centralized or decentralized? The environment may vary along two dimensions: complexity and predictability. The number of variables and the interrelatedness of the variables determine complexity. Predictability is determined by the available knowledge about the variables.

This research uses the goal orientation—effectiveness, efficiency, or both—and resulting strategies to discuss the relative focus on effectiveness among IT community stakeholders. This focus is used in our assessment in Chapter Three and discussed in greater detail in Chapter Four.

Factors to Consider in Assessing Occupational Specialties

From the theory and literature, we drew certain observations to use in our assessment of how various occupational specialties are affected by stakeholders:

- We define stakeholders as parties (individuals and groups) that have an impact on MPT decisions for a specialty or that are affected by such decisions. These stakeholders can be further characterized by power (money), legitimacy (OPNAV Instructions), and urgency (need).
- MPT systems that are high in distinctiveness, consistency, and consensus should provide organizational performance through stakeholders behaving appropriately. Stakeholders with competing goals that are not reconciled will have different strategies.
- Levels of training management need to be considered: Each of the stakeholders can be characterized by their "fit" within management levels.
- There should be an appropriate level of decentralization; decisionmaking should not be decentralized to the point where stovepiping is a problem.
- MPT decisions need to have clear lines of responsibility for decisionmaking. Authorities of stakeholders need to be known.
- Decisionmaking needs to be based on complete information; efficiency and effectiveness information needs to be shared across many levels in the organization.
- Organizational fit results largely from environment, goal orientation, and strategy. Maximizing both effectiveness and efficiency, while possible, is extremely difficult. Most organizations that prioritize efficiency do so at the expense of effectiveness. Knowing other stakeholders' goal orientation leads to greater understanding of the pressures and forces compelling disparate decisions.
- Change is difficult whether imposed through forces in the external environment, proposed as innovations/modernization, or limited through structural and cultural inertia. There must be enough initial and continuing change-based momentum to overcome inertia.

CHAPTER THREE

Assessing Ratings

In this chapter, we identify the stakeholders involved in selected personnel processes for IT and three comparison communities—hospital corpsman (HM), firecontrolman (FC), and the AF specialty cyber systems operations. The comparison ratings were selected for their similarities to and differences from the IT rating, described in the context of the discussions below. These processes were selected both because they are of interest to the research sponsor and also because they are key processes affecting the performance of these communities. Of interest to the research sponsor is the relationships among stakeholders. We have included those stakeholders with the greatest power, legitimacy, and urgency; so not all identified stakeholders will be referenced.

Is there an optimal structure of relationships among stakeholders for these processes? Our review of the literature, service directives, and interviews led us to assume a set of conditions that an optimal stakeholder set of responsibilities would meet. We will assess the four communities against these factors following a description of each community and its stakeholder relationships.

Navy Rating Characteristics and Distribution

Do the demographics and characteristics of the rating make a difference in terms of training processes or outcomes? We compared data for the Navy ratings in terms of size (billets authorized), Navy enlisted classification (NEC) composition, and units and commands where used.

Size and NEC Composition

Including those in training, the HM rating is by far the largest, with about 26,700 billets authorized. Of these, about 20,120 (75 percent) have a requirement for one of more formal NECs beyond the initial "0000," which designates an apprentice (informally called Quad 0). The IT rating is next with about 9,910 authorizations, and FC Aegis is smallest with about 6,630 billets authorized. Of the 9,910 IT billets, 3,006 (30 percent) have an authorization for at least one NEC, as does 2,411 (36 percent) of the 6,630 authorizations for FC.

In the HM rating, one NEC (8404 Field Medical Service Technician) accounts for 47 percent of all billets authorized, and no other NEC accounts for more than 7 percent of authorizations. In the IT rating, the distribution is more widespread, with five NECs having more than 10 percent of authorizations.[1] The largest is 2735 (Information Systems Administrator), with 25 percent of authorizations. Other large ones are 2780 (Network Security Vulnerability Technician) and 2379 (Transmission System Technician), with 17 and 12 percent, respectively. The FC rating is even more dispersed by NEC, with no one NEC having more than 6 percent of authorizations.

Units and Commands

HM authorizations exist in 23 major commands, compared with 26 for IT authorizations and 19 for FC authorizations. Each rating is clustered in certain, but different, Navy major commands. Not surprisingly, the HM rating is used primarily in the clinics and hospitals of the Navy Bureau of Medicine and Surgery (BUMED), with the second largest group used by the Commandant of the Marine Corps (CMC) in units in their ground combat element, with NEC 8404. IT personnel are primarily used in the U.S. Fleet Forces Command (USFFC) and the Pacific Fleet. The difference is that certain shore establishment organizations, such as computer and telecommunications stations and information operations commands, are part of USFFC and account for about 1,800 authorizations at the NEC level of detail. The FC rating is also primarily used in these same two commands. See Table 3.1 for a list of percentages of HM, IT, and FC ratings in each major command.

At the Unit Identification Code–level of organization, HM personnel are used in 1,673 different units, while IT personnel are used in 1,167, and FC personnel in 435. Each of these units averages 16 HM personnel, 15 FC personnel, and 9 IT personnel. If we arbitrarily look at units with 30 or more authorizations, for the HM rating, only 10 percent of all units have 30 or more, but this total accounts for 79 percent of all authorizations. The primary users are carriers and military health facilities on shore. For the IT rating, only 5 percent of these units have 30 or more, and these units account for 43 percent of all authorizations. Primary Unit Identification Code–level unit users are carriers and the shore establishment. For the FC rating, a larger number of units, 22 percent, have 30 or more authorizations, and these units account for 75

Table 3.1
Major Command Authorizations (in percentage)

Major Command	HM Rating	IT Rating	FC Rating
BUMED	60		
CMC	22		
USFFC	6	50	33
Pacific Fleet	4	28	38
SPECWARCOM		5	
Office of Naval Intelligence		3	
NETC			15
Chief of Naval Personnel			6

[1] We recognize that the IT rating is undergoing change at the NEC level of detail. The data we use have only very small numbers of the new NEC.

percent of all authorizations. Primary users are destroyers (DDGs), cruisers (CGs), and carriers (CVNs). In sum, the HM rating appears to be most heavily concentrated and clustered in the shore-based units of the BUMED military health system. The FC rating is largely in the major sea commands and mainly associated with units of the surface Navy. While IT personnel are used in the major sea commands, they are more widely dispersed, with a large number in the shore establishment of those commands.

Summary

These ratings have different sizes and distributions at the NEC- and unit-level of detail. The HM rating has one dominant user and one dominant NEC, and over 70 percent of its authorizations require the advanced knowledge or experience of a formal NEC beyond the apprentice "Quad 0." The IT rating is spread among a large number of users on both ship and shore, with no dominant NEC. Only 30 percent of IT authorizations require the advanced knowledge or experience of a formal NEC. The FC rating is largely sea and combat-system based and is widely dispersed by unit and NEC, with 36 percent of authorizations requiring a formal NEC.

Information System Technician

Information System Technician is an information technology enlisted rating. The Navy description of this rating states that IT personnel

> operate and maintain the Navy's global satellite telecommunications systems, mainframe computers, local and wide area networks, and micro-computer systems used in the fleet. Administrative support is also provided with the operation of automated equipment that keeps records of personnel training, disbursement, health, assignments and promotions within the Navy. They ensure the all-important communications link between units at sea and stations ashore (Department of the Navy, Navy Personnel Command, 2011a).

The IT Rating Stakeholders

A Navy IT rating stakeholder is a party (individual or group) that has an impact (influence) on *manpower, personnel, and training decisions* for IT personnel or is impacted by such decisions. Listed below are some of the stakeholders (including the processes in which they are involved):

- OPNAV N1 (accessing, sustainment training, and assigning)
- IT community manager (accessing)
- NPC (accessing and assigning)
- Naval Recruiting Command (accessing)
- OPNAV N2/6 (initial and interim training for new equipment)
- SPAWAR (initial and interim training for new equipment)
- PEO C4I (initial and interim training for new equipment)
- Center for Information Dominance, NETC (sustainment training)
- Naval Air Forces (performance)
- Naval Surface Forces (performance)
- Submarine Forces (performance)
- Navy Cyber Forces (performance).

See Appendix A for more details about stakeholder roles and responsibilities.

Selected Personnel Processes for the IT Rating

Figure 3.1 illustrates the personnel processes selected for discussion: accessing, training, and assigning. The resulting performance is also indicated in this figure, as are the stakeholders that affect these processes and that are affected by the resulting performance of IT personnel.

As the figure indicates, N1 funds the accession of IT personnel, the community manager at NPC determines the accession numbers and prerequisite qualifications, and the Naval Recruiting Command recruits the new entrants.

Training for IT personnel is divided into initial training for new equipment and sustainment training (training steady state), which are developed and funded differently from one another. As new equipment is developed, the PEO, as training support agent, is responsible for identifying and resourcing the training required to support the new equipment. SPAWAR is a large organization supporting the PEO and providing the expertise for these training assessments. N2/N6, as the resource sponsor, is indirectly funding the training and also has substantive input both for the new equipment and also for the supporting training. N1, specifically N15 (Education and Training Division of N1), also has input into the new equipment training, as part of its participation in the Navy Training System Plan (NTSP) approval pro-

Figure 3.1
The IT Rating Processes and Stakeholders

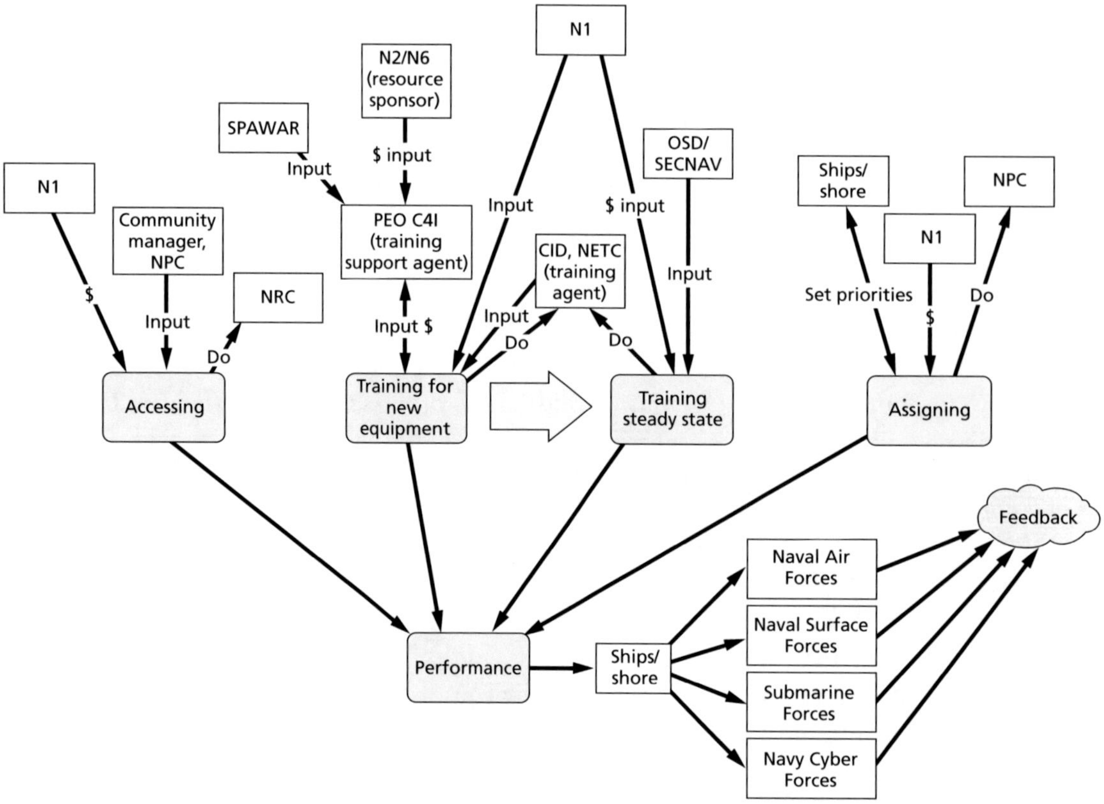

NOTES: In this and the following series of figures, "$" represents funding, "Input" is policy or program direction to include numerical goals, and "Do" represents execution of the process.

RAND *TR1122-3.1*

cess.[2] This participation is important because once the training is developed and the systems are delivered, this training ceases to be considered training for new equipment. All training is then incorporated into the A school and/or C school (occupation-specific training) as sustainment (steady-state) training for IT personnel, and N1 controls the curriculum (N15) and funds the individuals account (N12) that pays for the students to attend the training. N1 is also the resource source for the schoolhouse and trainers. The Office of the Secretary of Defense (OSD) and SECNAV can have input regarding the content of training through the Chief Information Officer (CIO), who levied a requirement for Information Assurance certifications that affected training content for IT personnel.

Assignment for IT personnel is handled consistently with the assignment for most Navy ratings, with Type Commanders (TYCOMs) for receiving units indicating their priorities for assignments, NPC facilitating the assignments, and N1 funding permanent change of station moves as well as the "individuals account," which includes transient manpower. While there are dysfunctional aspects of the assignment process, discussed in greater detail in prior research (Thie et al., 2009), these problems are not specific to IT personnel. It is worth noting, however, that while the receiving units can prioritize billets to be filled (or not filled), they have no independent opportunity to adjust the personnel requested, for example, to request a more experienced IT in place of two junior IT personnel.

The performance of IT personnel depends upon the accession, training, and assigning practices and affects the receiving units, those ships and shore establishments to which IT personnel are assigned. However, those users have only an indirect opportunity to provide feedback regarding IT personnel performance. IT personnel are assigned in small numbers to many units, ships and shore. Further, their feedback mechanism is through at least four different type commands (Naval Air Forces, Naval Surface Forces, Submarine Forces, and Navy Cyber Forces), spread between Pacific Command and USFFC, that have a mix of ship and shore responsibilities.

IT personnel training is a special focus of our research, since improved training is generally accepted as improving performance and increasing effectiveness. The preceding research (Thie et al., 2009) described training changes that were being considered for IT personnel and also analyzed the effect of converting all IT personnel to six-year enlistees, with the benefit of attending both A school and C school prior to assignment. That work concluded that effectiveness gains were possible if all IT personnel attended C school. While there were short-term costs to those changes, there were potential long-term savings. Further, even the short-term costs could be ameliorated with the effectiveness gains resulting from more effective IT personnel. While A school for IT personnel has been lengthened, only a minority of IT sailors will attend C school prior to unit assignment. In essence, the rating lacks a periodic "modernization" review such as exists for hardware through the various acquisition PEOs.

Navy instructions provide insights regarding the difficulty of increasing the training for IT personnel. There were formally OPNAV instructions addressing the objectives of training (Department of the Navy, Chief of Naval Operations, 1989) and the cyclical review of shore-based training (Department of the Navy, Chief of Naval Operations, 1995). However, those instructions are now defunct and only lower-level stovepiped instructions stand in their place. For example, NETC Instruction 1500.8, *Human Performance Requirements Review (HPRR)*

---

2 See Appendix A for a description of this process.

(Department of the Navy, Commander, Naval Education and Training Command, 2010a) requires the learning centers to review individual training, with a focus on validated fleet requirements. Such reviews are required no less than every 36 months. There are also other triggers for training changes, such as revisions to an enlisted occupational standard, a ratings merger, or fleet feedback (see Department of the Navy, Commander, Naval Education and Training Command, 2010b). However, in the case of IT personnel, there is not a single enterprise or dominant stakeholder to fund training revisions based on such reviews. This omission is underscored by a 2008 memorandum from the Deputy Chief of Naval Operations (Manpower, Personnel, Education and Training) (N1) that establishes that PERS-4, which is part of the Navy Personnel Command, division directors and enterprises will collaborate on training priorities, but that no new courses or modified courses will proceed without "enduring offsets" or a "successful" program objective memorandum (POM) submission. Thus, in the absence of a higher-level instruction both requiring the process and establishing responsibility for the process, training is not systematically reviewed and improved for the IT rating.[3] This is in part because IT personnel are distributed across many units and many different types of units, and thus IT personnel lack a strong sponsorship from any one enterprise. Type commanders (TYCOMs), Naval Network Warfare Command (NAVNETWARCOM), and now Navy Cyber Forces appear not to have the influence to effect change in training for the IT rating by themselves. Moreover, none of them is the resource sponsor. The resource sponsor, N2/N6, has not professed a willingness to fund training improvements for IT personnel at the expense of other priorities for funding.[4] Lastly, our interviews indicate that neither the prior enterprise, NNFE, nor the present one, NIDE, plays a central role in these issues.

Navy Aegis Firecontrolman

Within this section, we identify the stakeholders involved in selected personnel processes for the FC rating. The Aegis FC rating was selected for comparison because it is an advanced electronics/computer field that requires relatively high entry scores for training aptitude, involves intensive training, and—like a subset of the IT rating—requires six years of commitment. The FC rating was an appealing comparison also because it has similar processes but a different learning center; a different resource sponsor; and a subset of the same IT rating's users, represented by type commanders. Moreover, the Aegis FC rating is undergoing changes in the training requirements needed by Aegis FC sailors, as CGs and DDGs that employ the Aegis weapon system transitions to Open Architecture. The training requirements for Aegis weapon system Open Architecture sailors are different than those of legacy CGs and DDGs, and effective approaches are needed to meet training needs.

FC personnel operate weapons systems on ships. These weapons systems include the Aegis weapon system, Tomahawk missile system, Close in Weapons System, and MK-86 five-inch gun. FC personnel receive training to maintain, repair, operate, and employ the surface ship weapons/weapons systems. FC personnel enter the Navy in the advanced electronics/computer field, which provides the extensive training in electronics needed to operate and maintain com-

[3] This could be true for other ratings as well.

[4] This information is based on interviews with N2/N6 personnel.

puter systems, radars, communications systems, and weapons fire control systems on board ships.

The FC Rating Stakeholders

Below is a list of FC rating stakeholders, with note of the processes in which they are involved:

- OPNAV N1 (accessing, sustainment training, and assigning)
- FC community manager (accessing)
- NPC (accessing and assigning)
- Naval Recruiting Command (accessing)
- OPNAV N86 (initial training for new equipment)
- SPAWAR (initial training for new equipment)
- PEO Integrated Warfare Systems (IWS) (initial training for new equipment)
- Program manager (PM) Aegis (initial training for new equipment)
- Center for Surface Combat Systems, NETC (sustainment training)
- Aegis Training and Readiness Center (sustainment training)
- Naval Surface Forces (performance).

Selected Personnel Processes for the FC Rating

Figure 3.2 illustrates the processes and stakeholders involved with FC training. The broad processes include the development of accession policies, training for new equipment, training for steady-state or sustainment training, and assignment of personnel. As for the IT rating, N1 funds the accession of FC personnel, the community manager at NPC determines the accession numbers and prerequisite qualifications, and the Naval Recruiting Command recruits the new entrants. Training for new equipment and steady-state training use the same processes, but all of the stakeholders involved in training for this rating are different with the exception of N1. With respect to the assigning process, the FC rating is mainly used on ships and not a mix of ship and shore, as with the IT rating. A significant difference from the IT rating is in the performance and feedback loop, where there is a single user and type command for the FC rating, while performance feedback is split among multiple users and type commands for the IT rating.

The stakeholders involved with FC personnel training are many. For accession policies, the community manager provides input with regard to the Armed Services Vocational Aptitude Battery (ASVAB) qualifying scores for FC personnel training.[5] The recruiting command accesses qualified applicants, and N1 funds the NEC-producing courses.

When new equipment is fielded to the fleet, the Aegis PM develops and oversees the development and planning of the installation. The PM funds the vendor training, curriculum development, and schoolhouse.

The Director for the Surface Warfare of Division of N8 (OPNAV N86) is the resource sponsor for surface warfare. NAVMAC supports the training for new equipment by participating in the job task analysis, where it is determined which personnel (junior, journeyman, or master) will perform which function; NETC and its subordinate learning center, the Center for Surface Combat Systems (CSCS), develop the curriculum, identifying shortfalls and gaps in the training due to new equipment and developing curriculum to address the gaps. The

---

[5] ASVAB helps to define levels of aptitude as a basis for serving in certain specialties.

Figure 3.2
The FC Rating Processes and Stakeholders

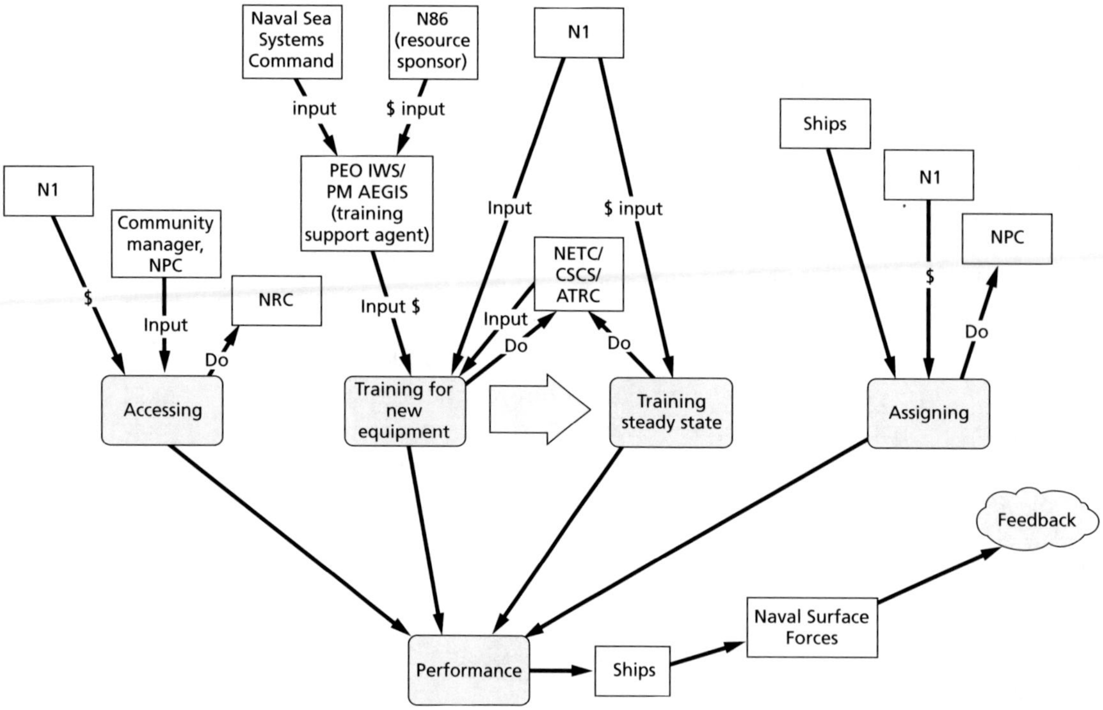

RAND *TR1122-3.2*

Aegis Training and Readiness Center (ATRC), part of CSCS, delivers FC training to new and current Aegis FC crewmembers.

At the ready for training (RFT) date, OPNAV N15 assumes the responsibility of funding sustainment training in the NETC learning centers. The office funds instructors and facility costs, including NEC-producing courses that constitute the bulk of Aegis FC training. NETC/CSCS reviews curriculum and provides input or makes curriculum changes to assure training effectiveness. ATRC executes delivery of sustainment training.

There is a collaborative approach to assignments. NPC assigns personnel to fleet units, and the ships address their manpower needs through the TYCOM and NPC, which generate personnel requisitions for training and assignment.

A training feedback loop provides input to training authorities on the needs to adjust requirements as necessary.

Navy Hospital Corpsman Rating

The enlisted health care HM rating is described by the Navy as follows:

> Hospital Corpsmen (HM) perform duties as assistants in the prevention and treatment of disease and injury and assist health care professionals in providing medical care to Navy people and their families. They may function as clinical or specialty technicians, medical administrative personnel and health care providers at medical treatment facilities. They also serve as battlefield corpsmen with the Marine Corps, rendering emergency medical treat-

ment to include initial treatment in a combat environment. Qualified hospital corpsmen may be assigned the responsibility of independent duty aboard ships and submarines; Fleet Marine Force, Special Forces and Seabee units; and at isolated duty stations where no medical officer is available (excerpted from the rating description in Department of the Navy, Navy Personnel Command, 2011b).

The HM rating has both similarities to and also differences from the IT rating. Both HM and IT ratings require high ASVAB scores and extensive training and result in skill sets that are easily transferable to civilian employment. Like IT personnel, many HM personnel attend only A school before serving in a duty station and return later for C school. Those are the key similarities that suggested a valid comparison. However, there are considerable differences between the stakeholder patterns evident in the HM rating, as compared to the IT rating, which make this a valuable comparison.

The HM Rating Stakeholders

Below is a list of the highly involved stakeholders for the HM rating. Next to each stakeholder is noted the process(es) with which they are involved. Of note is that BUMED is involved throughout these processes, from determining entry qualifications to assessing performance.

- N1 (accessing, training, and assigning)
- HM rating community manager (accessing)
- Naval Recruiting Command (accessing)
- BUMED (accessing, training, assigning, and performance)
- NPC (assigning)
- Fleet (performance)
- Marine Corps (performance).

Selected Personnel Processes for the HM Rating

Figure 3.3 illustrates the key stakeholders for selected personnel processes in the HM rating.

The accession processes are somewhat similar to those of IT personnel, as N1 funds the accession of HM personnel, the community managers provide the substantive input regarding numbers and qualifications, and the Naval Recruiting Command actually accesses HM personnel. However, BUMED, the user, has input into this process and the policies as well, with formal authority as the resource coordinator for Navy medical requirements. In this role, the Surgeon General coordinates with N1 for medical manpower requirements. Two of the three Enlisted Technical Advisors, the enlisted personnel who work with the community managers, serve in billets that are funded by BUMED, and BUMED provides considerable input through these Enlisted Technical Advisors.

The training for the HM rating is different from that for the IT rating in that there is essentially no new equipment training. Medical equipment is purchased commercial off the shelf, and so there is no development of unique new training for this equipment. Further, although the schoolhouse receives new equipment as part of the purchase contract, all newly purchased equipment is delivered to the clinics and hospitals at the same time, or slightly before, the equipment delivery to the schoolhouse. Thus, the schoolhouse need not train HM personnel on old systems while also training for new systems. When new equipment, such as a new X-ray machine, is delivered to a hospital, the purchase contract includes unit training for

Figure 3.3
The HM Rating Processes and Stakeholders

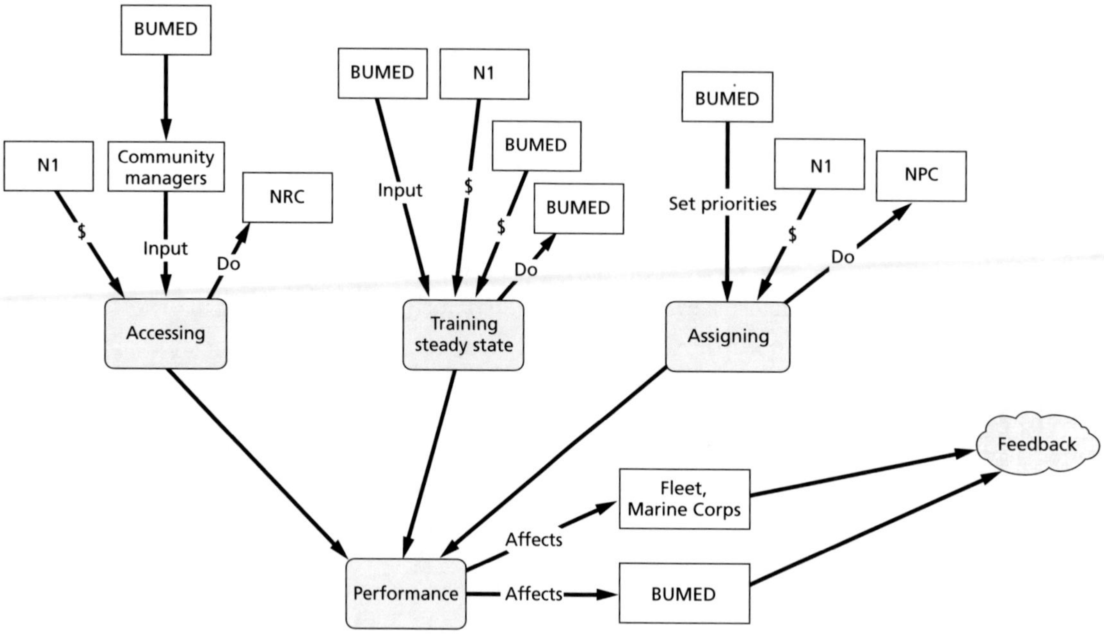

RAND *TR1122-3.3*

the key hospital staff. The steady-state HM personnel training is revised at that time to incorporate the new equipment. However, significant technology shifts that would suggest a major shift or accommodation in training, such as that seen as a result of the CANES program, are rare and unlikely. If there were such an instance, the training would still likely have been developed in the commercial world and simply adapted to and incorporated into the HM rating steady-state training. Another key difference in the training is that both N1 and BUMED pay for training. New HM personnel who have not yet served in a duty station are funded by the individuals account, but HM personnel who return to C school after serving in a duty station are funded by BUMED. BUMED funds training authorizations each year. Additionally, the schoolhouse and the instructors are funded with medical dollars, and the Surgeon General oversees this training.[6] The assignment process is also slightly different from that for IT personnel. Although detailers at NPC make the actual assignments, BUMED is involved in the prioritization of the billets because the majority of the billets are within BUMED organizations. All fleet and Marine Corps authorized billets (non-BUMED billets) receive the highest-priority fill. Following that, BUMED prioritizes the remaining billets in its organizations. BUMED has also determined the requirements and has sized its inventory to meet the overall requirements.

Although not apparent in the figure, another difference in the assignment pattern of the HM and IT ratings is that most HM personnel are working in BUMED environments, with

---

[6] Some of this funding comes directly from OSD, Health Affairs, and some of it comes through BUMED from OSD, Health Affairs. As of 2011, enlisted medical training for the Army, Navy, and Air Force is being consolidated at Fort Sam Houston. The Navy Surgeon General, not NETC, is responsible for this medical training.

other HM personnel and medical professionals.[7] This has implications for the feedback loop. While BUMED actively seeks feedback from the fleet and Marine Corps about the skill sets and performance of HM personnel, it also actively tracks and maintains occupational standards for the HM rating in BUMED organizations. This is done by the curriculum management personnel at the Navy Medicine Manpower, Personnel, Training and Education Command within BUMED, who observe and update the skill sets necessary for the HM rating and update the training as necessary. There is a consistent and continuous modernization function. BUMED also has the capability to adjust its requirements for HM personnel if it determines that a particular BUMED organization would benefit from having fewer HM personnel that are more highly trained or at higher pay grades rather than a greater number of more-junior or less-trained HM personnel.[8]

As another example of changes to the rating that BUMED is interested in and capable of making, BUMED recently increased the prerequisite ASVAB requirement for HM personnel. It did so after extensive study by Navy Studies, Research, and Technology to assess the effects of the change. The result has been a reduction in A school nongraduation from approximately 30 percent to approximately 5 percent.

In summary, there are significant differences between the IT and HM ratings. In particular, for the HM rating, BUMED is involved in every process from entry to assigning. Moreover, it is the primary user and does not have an intervening TYCOM for the feedback loop.

Air Force Cyber Systems Operations

The AF recently revised its cyberspace support career field (similar to Navy ratings) by merging three former career fields: communications—electronics, knowledge operations, and communications—computer systems. Eleven new specialty codes were created including one for *cyber systems operations*. This specialty supervises and performs cyber systems operations and executes associated information systems support programs, both in garrison and at deployed locations. The personnel perform system administration on C4I and various functional area platforms. As such, it (and the other ten specialties in this field), while slightly broader, is the counterpart of the multiple Navy NEC in the IT rating area, such as information systems administrator.

This section first describes the AF development system before illustrating specific personnel processes. Differences between the Navy IT rating and the AF cyber systems operations specialty are partly based on differences in procedures.

Cyber Systems Operations Force Development in General

The AF has been revising its personnel and training processes in recent years and currently focuses on an overall force development construct. As part of this process, Headquarters AF (the Secretariat and Air Staff) appoints career field managers to ensure development, implementation, and maintenance of career field education and training plans. The focus is on force

---

[7] This proportion increases during peacetime, when the HM personnel who surge to serve with the Marine Corps during wartime deployments (and some HM personnel who serve with the fleet during wartime) are instead assigned to clinics and hospitals. These wartime assignments are referred to as Marine Corps HSAP and Fleet HSAP.

[8] Assuming that these changes do not create an insupportable inventory.

development, defined as a series of experiences and challenges provided through education, training, and experience.

The AF executes force development through designated roles and responsibilities that cascade down from senior levels. Functional authorities (FA) are designated general officers and Senior Executive Service civilians serving as deputy chiefs of staff or Assistant Secretaries to provide oversight and advisory services related to functional communities.[9] The three-star general, Chief of Warfighting Integration and Chief Information Officer (SAF/CIO A6), serves as FA for this AFSC.[10] Among the responsibilities of the FA is to provide strategic oversight to include determination and prioritization of functional community requirements and to review career field health. These functional authority responsibilities are implemented through functional managers (FMs) and career field managers (CFMs).

FMs provide day-to-day management responsibility over several functional specialties and are responsible for ensuring their specialties are equipped, developed, and sustained to provide AF capabilities. The CFM is the AF focal point for each specialty and serves as the primary advocate for the career field. This person is appointed by the functional authority and must be a colonel for officer specialties and a chief master sergeant for enlisted airmen. There is no comparable functional responsibility chain from senior officer to a functionally oriented career field manager that we observed in the Navy. Each major command (MAJCOM) that uses the specialty also appoints a MAJCOM functional manager (MFM) for the specialty. Of interest, and different from the Navy, is that all of these responsibilities are embedded in the functional area and not in the personnel [Air Force Deputy Chief of Staff for Manpower and Personnel (A1)] or training (Air Education and Training Command [AETC]) bureaucracies. There is a "chain" of career field functional "czars" from the three-star level to the central functional coordinator, the career field manager. The AF philosophy is that "customers" should be in charge of training requirements. In this respect, it is more like the Navy HM rating.

The career field manager is the key link between the MAJCOM functional manager and AETC training pipeline managers (TPMs) at AETC headquarters and training managers (TMs) at the schoolhouses and communicates directly with them to disseminate AF and career field policies and program requirements. The TPMs are personnel assigned to HQ AETC and are responsible for cradle-to-grave management of initial skills formal courses.[11] They perform MAJCOM staff-level training management. CFMs and TPMs are in regular contact and talk weekly if not daily. CFMs are also in close contact with MAJCOM functional managers, who deal with training and other issues. If changes are needed, the CFMs will typically work with the lead MAJCOM or the MAJCOM with the most students for funding of the new requirement.

The TMs are at the geographic training groups (schoolhouses) and are responsible for training development, implementation, and management. The CFMs ensure development, implementation, and maintenance of the Career Field Education and Training Plan (CFETP) (U.S. Air Force, 2009) and serve as the chairperson for the Utilization and Training Workshop (U&TW) that is a periodic forum to determine and manage career field education and training requirements and resources. CFMs also assist AETC training managers and course super-

---

[9] These designations are formally contained in U.S. Air Force, 2008.

[10] Note that the naval CIO is not explicitly involved in IT rating management.

[11] AETC in the AF is a four-star AF Major Command and plays an important role throughout the training process.

visors with planning, developing, implementing, and maintaining all AFSC-specific training courses.

The AF uses the name *formal training* for what we have termed training steady state or sustainment training and *special training* for what we have termed training for new equipment. For formal training, the career field manager is largely in charge and determines specialty training tasks, requirements, and resources. The CFM also conducts periodic Utilization and Training Workshops and Specialty Training Requirements Team meetings (described below), which are supported by AETC training pipeline managers and include hosting, maintaining schedules, and providing the CFM with information to identify and justify unfunded training resource requirements.

The AF uses special training (typically done by contract) when regular training programs cannot satisfy the unique training requirements of a new system. As the single manager for special training, AETC determines whether government or contractor training best serves the needs of the AF. The acquisition agency (AF Materiel Command and PEOs) funds special training for new systems; A1 (Chief of Learning Division) oversees policy and planning for special training; and the Deputy Assistant Secretary for Budget in the Secretariat Financial Management organization supervises special training funds, which include budgeting, allocating, and accounting tasks. AETC assists the training planning teams to develop the Systems Training Plan and to set allocations to support system-specific training requirements. Moreover, AETC validates requests for special training, determines the most economical and efficient methods to meet needs and requirements, and determines whether sufficient funds are available to support the training if contracting is required. Typically a user command will be appointed as the lead command for a new system and has overall responsibility to ensure training is provided.

The AF uses a periodically scheduled workshop to formally assess training plans and resources. This workshop serves as a modernization mechanism and a quality control review. The career field managers and the AETC training pipeline managers (Headquarters Staff at AETC, distinct from the training managers who develop, implement, and manage training), who provide life-cycle management for formal training courses, determine the need for the Utilization and Training Workshops. Typically, one workshop is held about every three years, but there could be a shorter or longer cycle for particular specialties. These workshops look for disconnects among equipment, occupational measurement surveys, career field education and training plans, training seats, course content, resources (funding), and personnel. The workshop itself plays out over about a six-month period to allow time for coordination, research, and information gathering from the MAJCOM FM. A preliminary group [the Specialty Training Requirements Team] meets four months prior to the workshop to determine training requirements. The AETC/TPMs then task the TM to provide training options and resource estimates. The U&TW is then convened as an executive decision meeting cochaired by the AF CFM and the AETC/TPMs to determine which organizations (MAJCOMs) will furnish resources and when. The workshop also recommends long-term resource requirements.[12] If needed, there is

---

[12] In 2001, RAND was asked by the AF to informally review the mechanism between functional managers and the training community. In that review, the *Navy Training Requirements Review* (Department of the Navy, Chief of Naval Operations, 1995) was highlighted as a useful model because a senior group (policy group) considered recommendations for training changes that required additional resources. Similar processes were then instituted in the AF and continue to this day. In the Navy, that OPNAV instruction and the training review process no longer exist.

an issue resolution meeting when unresolved issues remain. A senior AETC official (usually AETC/A3T) chairs this meeting with participants from the Air Staff and MAJCOMs who have the authority to transfer resources to the training community or to accept a reduced level of training. Potential outcomes are reallocation of resources, reduction in training demands, or deferment of training changes until required resources can be provided through the corporate budgeting process.[13] In the AF, MAJCOMs fund formal technical training through a brokered PPBE process that moves needed resources to AETC.

Cyber Systems Operations Stakeholders

Below is a list of highly involved stakeholders for cyber systems operations. Next to each stakeholder is noted the process(es) with which they are involved. Of note is that a functional manager is involved throughout these processes, from determining entry qualifications to assessing performance.

- AETC (accessing, special training, and formal training)
- Functional authority (FA) (accessing and formal training)
- Functional manager (FM) (accessing and formal training)
- A1 (accessing, special training, and assigning)
- AF recruiting service (accessions)
- CFM (accessions, formal training, and performance evaluation)
- AF Materiel Command (AFMC)/PEO (special training)
- Deputy Assistant Secretary Air Force (DASAF) budget (special training)
- AETC/TPMs (formal training)
- MAJCOM FM (formal training and performance evaluation
- AF Personnel Command (assigning).

Selected Cyber Systems Operations Personnel Processes

Figure 3.4 illustrates the personnel processes for the AF specialty, and the processes are contrasted with the Navy occupations in the discussion. The differences from the IT rating are significant. The user community, represented in the AF by the functional authority and the career field manager, is directly involved in accessing, training, and the performance feedback loop. Also, rather than funding being provided by the OPNAV staff (as it is for the IT rating), training funding is provided through an operational four-star major command. Other AF MAJCOMs can also choose to provide additional resources to achieve their training needs, thus increasing immediate readiness (effectiveness). The processes for this AFSC have more in common with the Navy HM rating than with Navy IT rating.

The AF Recruiting Service, a subcommand of the four-star Air Education and Training Command, funds accessions. Policy and program information is provided from multiple sources, including A1 and the functional career field manager. AETC plays a consistent role in entering and training personnel. As indicated above, AETC has a critical role in new equip-

[13] Thus, there are three sources of specialty training funds. For long-term needs, AETC programs and/or budgets the funding through the AF program and budget process, which directly resources AETC as a MAJCOM. For immediate needs, AETC can first rebalance resources across all of its learning centers to optimize priority AF needs. If that is insufficient, a MAJCOM, as part of the UFT&W process, can choose to fund the needed training from its own resources. If the MAJCOM decides not to do this, then it knows the tradeoff it has chosen.

Figure 3.4
Cyber Systems Operations Processes and Stakeholders

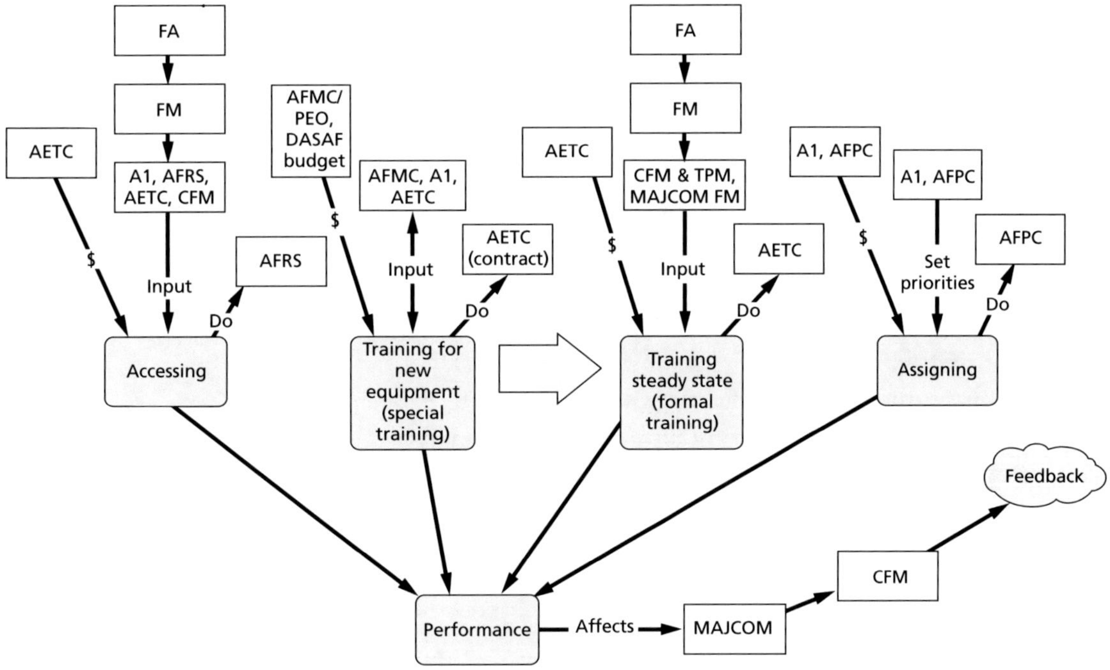

RAND *TR1122-3.4*

ment training, both by assuring the developer has usable training requirements and by funding (and later by providing the training, usually through a contract).

With respect to steady-state training, the functional career field manager, the AETC training pipeline manager, and the major command functional manager adjudicate requirements. These requirements are programmatically funded by AETC or by the major command if needs are more immediate.

Assigning processes for cyber systems operations is not dissimilar to all of the Navy processes. In terms of feedback, the AF processes explicitly provide for interactions between the MAJCOM, the training pipeline manager, the functional authority, and the career field manager.

Observations

Some general observations emerge when we compare the four figures. For example, accessing and assigning processes are generally similar with some minor differences. For the IT rating, there are more stakeholders, especially at the user level. For the FC rating, there is one dominant user and a resource sponsor that has close relationships with the surface community that uses the FC rating. For the HM rating, BUMED plays a role across the board, thus providing consistency in decisionmaking and in the ability to conduct tradeoffs. For the AF, the career field manager and the AETC play a role throughout the processes, with the user (the MAJCOM) significantly involved.

More specifically, our review of the literature and of three Navy ratings and one AF specialty code led us to the observations below. We recognize that the observations are based on a limited sample, and we make no claims that they are generalizable across all occupations. Moreover, in terms of favorable program (e.g., manpower, personnel, and training) outcomes based on practices and policies for these occupations, we rely on interviews and our own assessments. There are differences in policy, practice, and outcomes across the four occupations, particularly within the Navy and between the Navy and the AF.

Different ratings/services have different stakeholder patterns and responsibilities. The theory reviewed earlier and the differences we observe in the several ratings suggest that certain structures and stakeholder patterns are preferable, if not optimal, because they better integrate the goals of effectiveness and efficiency. For example, theory suggests the need for understanding the issues of effectiveness and efficiency when making decisions and the need for having clear responsibility ("ownership") for outcomes.

Even this limited sample illustrates significant differences within the Navy and between the Navy and the AF. In our scoring system, indicated in Table 3.2, a check mark represents at least partial agreement with the statement. Some of these observations are for conditions out-

Table 3.2
Observations Regarding the Four Communities

Observation	IT	FC	HM	Air Force Cyber Systems Operations
With Respect to Power				
Focus on efficiency and effectiveness, with ability to make tradeoffs between the two			√	√
The ultimate decisionmakers: users, rather than program and budget providers			√	√
With Respect to Legitimacy				
Training emphasized, with an identified functional lead, regular reviews, and hard metrics			√	√
A clear "ownership" of the specialty, with ultimate authority and central decisionmaking for policy and directives related to the community			√	√
Considerable influence in training and budgeting decisions by those with expertise in the career field			√	√
Fewer stakeholders involved; easy to gain consensus			√	√
A smooth transition of new equipment training into steady-state training			√	√
With Respect to Urgency				
Specialty concentrated: assigned to units in large numbers		√	√	Not observed
Central involvement in decisionmaking for the occupation by the units that benefit from performance improvements or suffer from decreases in performance			√	√
Performance prioritized		√	√	√

side the control of the manpower, personnel, and training communities (e.g., where and how the occupation is used). Most of the observations though should be within the control of these communities. Most notable is the absence of any check mark for the IT community.

Noted Is a Focus on Efficiency and Effectiveness, with the Ability to Make Tradeoffs Between the Two

The AF cyber operations community and the Navy medical community are able to balance effectiveness and efficiency. In the AF, the MAJCOM can fund the revision or expansion of individual training with MAJCOM resources. If a MAJCOM chooses not to fund training, the training does not occur, but the MAJCOM knows the impact of this decision. The decision to allocate funding or not to allocate funding for training suggests a decision between efficiency (cost savings) and effectiveness (enhanced performance). Likewise, BUMED has the ability to prioritize performance by changing requirements to result in fewer people of greater experience and training. In contrast, in the IT and FC ratings, decisions are based on efficiency criteria. For example, the training decisions resulting from the Information Technician of the Future program were based on near-term cost limitations.

Users, Rather Than Program and Budget Providers, Are Decisionmakers with Respect to Training

In the HM rating and in the AF, the user, rather than the resource provider, can make final decisions. In the HM rating, the user and the resource provider are the same entity. In the AF, the functional community plays a dominant role in the training process, and the major command that uses cyber systems operations personnel can choose to add resources for training. If the MAJCOM has sufficiently urgent needs, it can make resource decisions and influence training change. For the FC and IT communities, the resourcer is the decisionmaker.

Training Is Emphasized, with an Identified Substantive Lead, Regular Reviews, and Hard Metrics

The AF conducts periodic reviews of training that are led by the users and trainers. They also conduct surveys of supervisors and occupational studies to suggest and support training revisions. These processes are institutionalized in a Career Field Education and Training Plan (CFETP) that is rigorously enforced. The most recent review instituted changes for AFSC cyber systems operations. BUMED also conducts regular reviews of the occupational skills needed by medical personnel and the training to support those skills. These tasks are done by the head of curriculum management at the Navy Medicine Manpower, Personnel, Training and Education Command (NAVMED MPT&E). In contrast, there is no regular review of either the FC or IT community training, despite a former OPNAV instruction that specified regular review of training.[14] The most recent evaluation of training for the IT rating, the "IT of the Future" assessment, is notable because it did not emerge from a regular process, and it resulted in suggested improvements for IT personnel training, but there continues to be a struggle to implement all the recommendations. The Navy also periodically conducts "Deep Dive" assessments, but they are not regularly scheduled, and there is no identified lead for any future IT personnel training Deep Dive. Recently, NETC published an internal directive

---

[14] Department of the Navy, Chief of Naval Operations, 1995, established general policy for the cyclical review of shore-based training.

for training reviews that it might conduct, which is based on criteria outlined in the directive (Department of the Navy, Commander, Naval Education and Training Command, 2010a).

"Ownership" of the Specialty Is Clear, with Ultimate Authority and Central Decisionmaking for Policy and Directives Related to the Community

In the case of both AF cyber systems operations and the Navy HM rating, there is clear ownership and leadership of the communities. The three-star Chief of Warfighting Integration and Chief Information Officer, Office of the Secretary of the AF, serves as the functional authority of the AF cyber systems operations community. There are also functional managers and MAJCOM functional managers. Of special importance is that these individuals are all in operational units or functional staffs, rather than in staff positions in manpower and personnel organizations. Likewise, in the case of the HM rating, BUMED retains all decisions regarding these individuals, with the ultimate decisionmaker being the Navy Surgeon General. The ownership of the FC and IT ratings is less clear. The Navy CIO has a strategic plan for information technology, but it stops short of asserting ownership of the IT rating.

Those with Expertise in the Career Fields Have Considerable Influence in Training and Budgeting Decisions

As Table 3.2 indicates, in both the HM and the AF cyber systems operations communities, individuals with expertise in the career fields exercise considerable influence in the training and budgeting decisions. In the HM community, medical experts within BUMED make key training and budgeting decisions regarding the HM community. In the AF, the functional authority, functional managers, career field managers, and major command functional managers are all involved in ensuring that the cyber systems operations career field is equipped, developed, and sustained. In the case of the Navy IT rating and the Navy FC rating, the resource sponsor and the PEO or program managers provide funding and input for new equipment training, while N15 funds steady-state training. While the schoolhouses (CID and CSCS), which include individuals with IT and FC rating expertise, provide input for decisions regarding steady-state training, senior representatives of the IT or FC community are not typically involved in these decisions. Some, but not all, of the IT TYCOMs may be involved in such decisionmaking but not to the extent as in the AF and HM communities.

When Fewer Stakeholders Are Involved, It Is Easy to Gain Consensus

The HM community clearly has fewer stakeholders involved in its decision process, as BUMED is involved in decisions pertaining to accessing, training, assigning, and subsequent performance. The AF cyber systems operations community also has fewer stakeholders than do the IT and FC communities. Not only are FAs and FMs representing the same organization, but also, the CFMs, TPMs, and MAJCOM FMs have an integrated central process that provides input to training decisions, so that they do not always act as separate stakeholders in the decision process. In contrast, the FC community has more stakeholders involved than either the AF cyber systems operations or the HM communities, and the IT community is further complicated by the different TYCOM users.

New Equipment Training Transitions Smoothly into Steady-State (Sustainment) Training

For the HM rating, new medical equipment is purchased commercial off the shelf and training for new medical equipment is purchased from the contractor selling the equipment. Navy HM and other medical personnel receive new equipment training at the time that equipment is installed in the hospital or clinic. As a result, there is no individual training for new equipment, and thus, new equipment training was not portrayed in Figure 3.3. The AF also purchases new equipment training for cyber systems operations personnel from the new equipment purveyor. Moreover, the AF picks a lead command for the new systems that provides oversight of whether new equipment training is done well. As a result, one of the user MAJCOMs has broad oversight.

New equipment training for major combat systems, such as the AEGIS system, is more focused. Required operational capabilities (ROCs) are very specific and are enforced for these systems. Also, USFFC is more likely to be involved in the ROCs and the enforcement of the ROCs for combat systems. The enterprises for air and sea are more coherent than the enterprise for information. In contrast, new equipment training for the IT community is complicated by two key factors. First, training is confounded by the number of ship and shore installations of the new equipment. As a result, individual training for legacy systems persists long after the first new system is installed, as sailors are still receiving assignments to ships and shore installations that use the legacy systems.

The involved stakeholders are also a factor confounding the transition from new equipment training to steady-state (sustainment) training. IT rating new equipment training is funded by the resource sponsor through the training support agent. However, steady-state training is funded by N15. There is no effective process to reconcile the steady-state training curriculum with the new equipment training curriculum. As a result, the resource sponsor and training support agent can develop extensive training that N15 inherits and must fund as steady-state training. Moreover, N15 may delay transitions to sustainment training if it is not ready to accept the training transition costs.

The Specialty Is Concentrated: Assigned to Units in Large Numbers

The user base of the HM community is clear: The majority of HM personnel are concentrated in large numbers and serve in Navy medical clinics and hospitals. Most of the remaining HM personnel serve primarily with the Marine Corps. On the other hand, IT personnel serve both on ships and in shore facilities, in relatively small numbers; there is no significant concentration of IT personnel serving together, with the exception of large deck ships and shore communications stations. FC personnel are predominantly in ships of the USFFC and PAC fleet.

The Units That Benefit from Performance Improvements or Suffer from Decreases in Performance Are Centrally Involved in Decisionmaking for the Occupation

The stakeholders that are affected by the performance of the communities are involved in the training decisions for AF cyber systems operations and the HM rating. In other words, those who make training decisions deal with the consequences. Specifically, the MAJCOM (and the units represented by the MAJCOM) is most affected by AF cyber systems operations community performance, and BUMED is the key performance stakeholder for the HM community. In both of these instances, these stakeholders are directly involved in training decisions for these personnel. Both BUMED and the AF MAJCOM can choose to fund the modification of training if they perceive the need. In the case of Navy FC and IT ratings, the ships and both ships and shore installations (respectively) are the performance stakeholders. Yet, they have only indirect feedback to the training decisions made for these communities. The TYCOMs can provide input to the schoolhouses and to other stakeholders, but they do not have direct input into training decisions, and they do not have the resources to modify the individual training provided for these communities. Moreover, IT personnel performance is spread over multiple TYCOMs for ship and shore, while FC personnel performance affects only one ship-oriented TYCOM.

Performance Prioritized

Three of the communities considered herein have been linked to mission success. FC personnel are considered important to the mission because they operate AEGIS, a combat system. HM personnel are recognized as a vital medical resource critical to saving lives in hospitals, clinics, and in the fleet and among Marines. The AF Chief of Staff has stated that "cyber operations reinforce and enable everything we do, from administrative functions to combat operations."[15] IT personnel performance has not been recognized or prioritized in the same way or to the same extent.

---

[15] In *Cyberspace—Shaping the New Frontier* (2010) a video introduced by Lieutenant General William Lord, the AF Chief Information Officer.

CHAPTER FOUR

Goal Orientation of IT Stakeholders

The prior chapter described the stakeholder situations for each of the communities considered in this report. One observable stakeholder difference between the IT and the comparison communities is that the stakeholder interactions are more complex for IT personnel; there are more stakeholders, and they are from different Navy organizations. As a result, it is useful to examine the goal orientation of these IT stakeholders. The following discussion explains the theory of goal orientation, describes a method for determining goal orientation, and then illustrates the goal orientation of the IT stakeholders.

Goal Orientation: Effectiveness, Efficiency, or Both?

Goals or objectives usually can be constructed in a hierarchy with lower-level objectives supporting higher-level goals. Ultimately, at the highest level, effectiveness and efficiency can emerge as two dimensions rather than the opposite ends of a single scale. Efficiency is a primary focus on inputs, use of resources, and costs. Effectiveness is a focus more on outputs—products or services and growth. These are relative choices that are not good or bad in themselves. Some organizations place a higher priority on efficiency, focusing on minimizing costs, while other organizations emphasize effectiveness, focusing on seizing leading-edge innovation irrespective of cost. All organizations value both effectiveness and efficiency, but is one more dominant than the other? (Burton, DeSanctis, and Obel, 2006.)

We used a series of diagnostic assessments based on our interviews to assess whether Navy stakeholder organizations were more aligned to effectiveness or efficiency or accorded both equal priority. A stakeholder that emphasizes efficiency focuses on utilization of the smallest amount of resources necessary to produce its products or services, continues to do what it has done in the past while refining those processes for continued improvement, and operates in a more stable environment. A stakeholder that emphasizes effectiveness focuses on its goals, takes less care to attend to the efficient use of resources, has a more volatile environment, and constantly develops new ideas to implement. A stakeholder that focuses on both effectiveness and efficiency equally typically has a competitive, complex, and volatile environment; innovates in products and services; and uses resources wisely (Burton, DeSanctis, and Obel, 2006).

Organizations are not necessarily locked into one of these dimensions over time. Some may be, but it is also possible that organizations that usually focus on inputs, use of resources, and costs are occasionally disrupted by revolutionary periods of change in which outputs, products or services, and growth are the focus. An organization that prioritizes both effectiveness and efficiency simultaneously focuses on inputs, resources, and costs on the one hand and outputs, products or services, and growth on the other. In recent years, the Navy has focused more on efficiency, as supported by the Balisle study, an independent effort commissioned by Admiral John Harvey and produced by a panel that includes flag officers. The Balisle report asserts,

> It appears the effort to derive efficiencies has overtaken our culture of effectiveness. . . . The material readiness of the surface force is well below acceptable levels to support reliable, sustained operations at sea and preserve ships to their full service life expectancy. Moreover, the present readiness trends are down.[1]

Strategies That Fit with a Goal Orientation

Strategy is an operational statement of an organization's objectives and should fit with it. A number of researchers have posited dimensions of strategy. Burton, DeSanctis, and Obel (2006) base their typology on Miles and Snow (1978) and characterize possible strategies in five ways. We adapted the strategy terminology for this effort and use the following: react, maintain, experiment, analyze without innovation, and analyze with innovation (see Table 4.1).[2]

Burton, DeSanctis, and Obel (2006) suggest the five strategy types "fit" well with certain goal orientations. The "reactor" has neither an efficiency nor an effectiveness goal. The "maintainer" has an efficiency goal. The "experimenter" has an effectiveness goal, while the "analyzer," with or without innovation, has both efficiency and effectiveness as goals.

These concepts are illustrated graphically in Figure 4.1, which places these strategy types on x- and y-axes, where the x-axis represents effectiveness and the y-axis represents efficiency.

For our research purposes, rather than suggesting a strategy to fit with a stakeholder's objective, we designed an assessment instrument that allows us to discern a stakeholder's strategy and then use that as a means to uncover basic objectives.

Goal Orientations of IT Stakeholders

In Figure 4.2, we chart the goal orientations of the IT stakeholders, having considered them against Burton, DeSanctis, and Obel's (2006) criteria. These evaluations are based on a subjective assessment discussed in greater detail in Appendix B.[3] In the figure, the horizontal x-axis

---

[1] Quoted in Ewing, 2010, p 1.

[2] Burton, DeSanctis, and Obel (2006) use "reactor," "defender," "prospector," "analyzer without innovation," and "analyzer with innovation."

[3] We drew 38 statements from the literature (largely Burton, DeSanctis, and Obel, 2006) that could be used to assess whether a stakeholder had an effectiveness or efficiency orientation. These statements were similar to those in Table 4.1. Three senior researchers involved in this research then scored each of these statements for selected stakeholders on a five-item scale from "completely disagree" to "completely agree." The basis for the assessment was interviews conducted by the researchers, written material, and prior research. Differences between researcher scores generally reflected a difference in emphasis (e.g., strongly agree versus agree). Such differences were discussed, and a group score for each item was agreed on. The individual item scores were then aggregated into an overall score for both the effectiveness and efficiency dimensions. The calculation method is discussed in more detail in Appendix B. While these evaluations are based on interview and other data, we recognize this process is subjective and other raters might have scored the items differently. Thus, we make no claim that the scores are "correct" in some fashion.

Table 4.1
Goal Orientation Strategies

React
No innovation
No systematic anticipation, plan, or look to the future
Reacts; tries to adjust to new situations
Pursues innovation without any focus
Makes decisions based on bad news arriving
Problems emerge as surprises; dealt with as they occur
Focus is on internal organization and bureaucratic struggles
Maintain
Emphasis on maintaining competitive position
Extrapolates the past as a basis for future planning
Works to fend off competitors
Has niche to protect
Cannot change much or change quickly
Strategy of doing the same thing efficiently
Emphasis on quality; seeks improvement in existing processes and existing products/services
Slow to make significant change
Experiment
Focuses on innovation of new things
Searches continually for new opportunities
Experiments regularly with new ideas, new technology, and new processes
Creator of change
Not much concerned with using its advantages or with developing efficiencies in its use of resources
Constantly questions the status quo
Can make large changes in products and services
Quality not primary concern
Leads in innovation; constant stream of new products/services
Has passive innovation strategy or a copy strategy
Analyze without innovation
Imitates others successes with products/services
Efficient in utilization of resources
Makes small changes in existing processes/products/services
Can change by following others
Adept at doing much the same thing efficiently with a few changes from time to time
Exploits current position of efficient resource utilization
Analyze with innovation
Develops new products, services, and their delivery processes
Purposely innovates and searches for new products and services
Surveys widely to look for opportunities
Innovates to meet customer needs
Invests in technology and capitalizes on results

Figure 4.1
Effectiveness and Efficiency Strategies

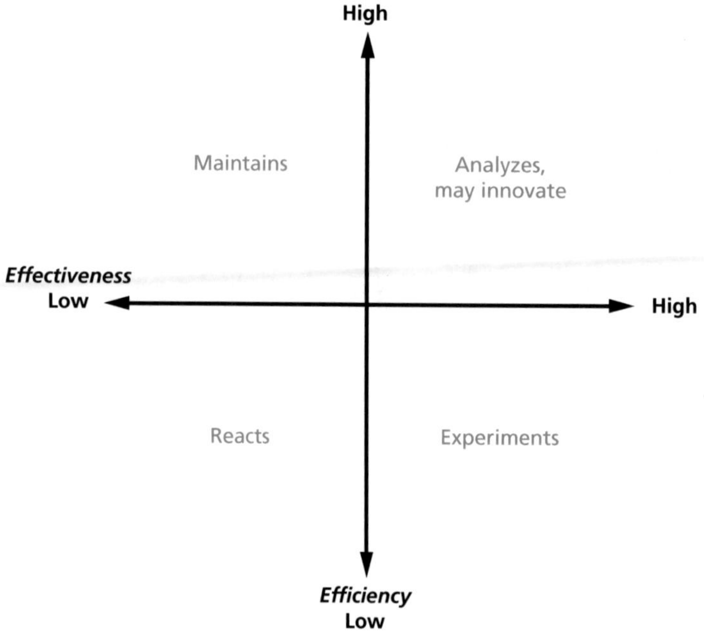

RAND *TR1122-4.1*

Figure 4.2
Goal Orientations of IT Stakeholders

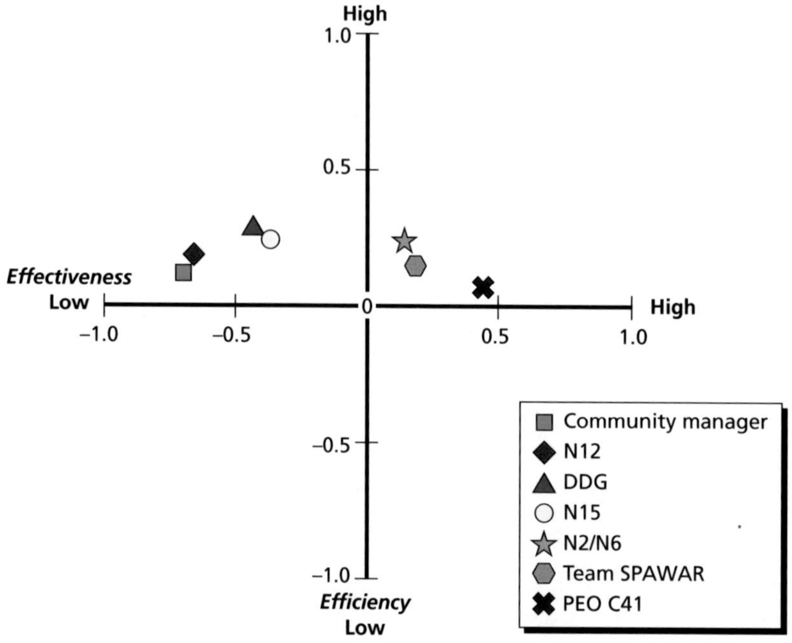

RAND *TR1122-4.2*

represents effectiveness, such that the right side of the chart indicates higher effectiveness orientation and the left side of the chart represents lower effectiveness orientation. The vertical y-axis represents efficiency. As the figure indicates, the analysis suggests that all of the organizations are efficiency oriented, in that all reside above the x-axis, or in the upper half of the efficiency axis. Most of the stakeholders in the upper left quadrant are MPT organizations. The DDG, a typical surface ship representing the users, is also placed in this quadrant because such users have only limited resources and no opportunity to make efficiency/effectiveness tradeoffs. Instead, the personnel and equipment communities have been compelling ships into an efficiency focus by reducing the resources available to them.

Of the three stakeholders who place greater emphasis on effectiveness, N2/N6, who is the primary bill payer for IT systems and new equipment training, values effectiveness but is constrained by real-world funding limitations. Only the PEO and Team SPAWAR, who have very forward-looking missions and fewer fiscal constraints, are able to adopt a more innovative analysis strategy. However, these stakeholders are minimally involved in the MPT decision processes for IT personnel, as they affect only the training for new equipment and are not the sole decisionmakers for that training.

In conclusion, while all of the stakeholders have a positive efficiency strategy, this illustrative analysis suggests that not all of them have an effectiveness strategy. The three stakeholders who focus largely on equipment modernization have both an efficiency and an effectiveness orientation. The other stakeholders, primarily those focused on MPT, have efficiency orientations. Indeed, the majority of the stakeholders prioritize efficiency over effectiveness, in the context of their "maintain" strategy. These differences indicate that the cluster of organizations focused less on effectiveness are better able to communicate within that cluster than across to the other cluster focused more on effectiveness. They both share the efficiency "gene" and should be able to meet on those grounds. However, too frequently it appears that even this is difficult as the clusters become focused on using the others' resources to preserve their own efficiencies. Effectiveness discussions are even more difficult. And there appears to be no central authority with legitimacy, power, or urgency to bridge the divide.

The efficiency emphasis of these Navy organizations is consistent with recent statements from the Secretary of Defense and the Chief of Naval Operations, who have underscored efficiency:

> Meeting real-world requirements. Doing right by our people. Reducing excess. Being more efficient. Squeezing costs. Setting priorities and sticking to them. Making tough choices. These are all things that we should do as a Department and as a military regardless of the time and circumstance. But they are more important than ever at a time of extreme fiscal duress, when budget pressures and scrutiny fall on all areas of government, including defense.
>
> —*Secretary Gates, January 6, 2011 (Department of Defense [DoD], 2011a)*

> The Navy enthusiastically participated in Department of Defense efficiency efforts . . . . These savings and changes will enable us to be the Navy the nation needs today and into the future.
>
> —*Admiral Roughhead, January 6, 2011 (DoD, 2011b)*

The predominant Navy emphasis is on efficiency, and the expenditures planned with the efficiency savings include primarily weapons systems rather than people (DoD, 2011a). Some of the efficiency savings will be spent on people, to increase the personnel on ships, because it was recognized that prior savings went too far. This is a typical organizational response from the Navy. Another strategy more focused on effectiveness would be to provide better and longer training by using the additional end strength to fund the individuals account for training. This would result in more productive and effective sailors, rather than simply more sailors.

Nonetheless, the modernization of military equipment has included inherent effectiveness and efficiency gains, which is evident in the smaller inventory of ships and airplanes in today's military. The weapons systems perform better than prior systems. The emphasis on effectiveness in weapons systems is not gone from the Navy, but there is an accompanying emphasis on efficiency. Thus, it might be reasonable in the Navy's type commanders to see a greater emphasis on effectiveness. However, the prior chapter discussed how those type commanders have only limited interest and diluted input to management decisions regarding the IT rating. Instead, the organizations that serve as predominant stakeholders in decisions regarding the IT rating have a more predominant emphasis on efficiency.

Also, an open question is whether the MPT communities shown in Figure 4.2 could move to the upper right quadrant, where it appears the Navy HM and the AF cyber systems operations communities are located. A modernization (analytical) process that leads to MPT change from time to time could be implemented. At one time, the Navy had such a process in place.

CHAPTER FIVE

Conclusion and Recommendations

Conclusion

This report addresses the stakeholders involved in the manpower personnel, training, and education decisions regarding the information technician rating. The focus of this report is on the implications of these different stakeholders on the cost and effectiveness of IT personnel. We define stakeholders as those parties (individuals or groups) that have an effect on MPT decisions or are affected by those decisions. Consistent with the literature, we focus on the stakeholders with legitimacy, urgency, and power. Theoretically, the system responds to stakeholders that have legitimacy, power, or urgency. Legitimacy comes from service directives/instructions; urgency comes from fleet need. Power in the DoD typically comes from money.

This study described the management processes of comparison ratings selected from within the Navy for both similarity to the IT rating and also for differences. An AF community similar to the IT rating was also selected to observe relevant management patterns from another service. In the Navy, the FC and IT ratings use similar processes but have different outcomes because of their stakeholder patterns. IT personnel are assigned to many different units represented within four different type commanders. As a result, there is no single user of the IT rating that acts as a predominant stakeholder who is willing to invest in the IT rating and effect change. Urgency has little impact on the IT rating MPT. The FC rating MPT processes are similar to those of the IT rating, but the predominant user, Naval Surface Forces, is able to indirectly affect change with feedback to the MPT processes. The stakeholder pattern for the HM community is considerably different from that of either the IT or the FC communities. BUMED, the primary user of the HM rating, is considerably involved in the decisions for accessing, training, and assigning HM personnel. The "czarship" of BUMED is able to consider effectiveness/efficiency priorities and make changes where necessary. Urgency, legitimacy, and power are consolidated in a useful way. The AF system is different from those in the Navy. While the AF management of cyber operations includes many stakeholders, there is (as for the HM rating) an identified dominant stakeholder responsible for the development and effectiveness of cyber operators. The functional authority is supported by functional managers and career field managers, all of whom are located in operational units, rather than in personnel staff units. Their participation, along with the active participation of the MAJCOM, ensures that the performance and effectiveness of cyber operators are emphasized in MPT decisions. Thus, while the FC rating, the HM rating, and AF cyber operations all have predominant stakeholders with legitimacy, power, and urgency, the IT rating does not have an identified predominant stakeholder, and IT rating users are involved only peripherally in MPT decisions.

The literature indicates that there should be an appropriate level of decentralization, but that decisionmaking should not be so decentralized so as to develop problematic stovepipes. In the case of the IT rating, there is no central corporate oversight of training decisions. Although there used to be an OPNAV instruction pertaining to centralized and periodic training review, now there are stovepiped instructions from multiple stakeholders that do not assign overall responsibility for ensuring that training improvements occur. These diffused responsibilities for training are dysfunctional for the IT rating and are made without a corporate or strategic perspective regarding tradeoffs between efficiency and effectiveness.

The literature also offers insights regarding the importance of organizational fit, which results from consistency within goal orientation (efficiency and/or effectiveness) strategy and environment. Given the complex stakeholder pattern for IT personnel, this research explored the organizational fit of IT rating stakeholders. Most of the IT rating stakeholders are efficiency driven and employ a "maintain" strategy. Only three IT rating stakeholders are both efficiency and effectiveness driven, and these stakeholders lack either the opportunity or the priorities necessary to effect change in MPT decisions, such as those regarding occupational revisions and steady-state/sustainment training.

This is an important point, because organizational change is difficult under the best circumstances; structural and cultural inertia will resist change. Thus, there must be both stakeholder action and identified triggers that will create change. In the case of AF cyber operations and the Navy HM rating, both exist. There are identified processes that evaluate the occupations and the training and initiate required changes on predetermined schedules. In the case of the IT and FC ratings, there is also an NETC instruction that identifies triggers to evaluate and improve training. However, a dominant stakeholder to sponsor and resource change is still required. The FC rating has that dominant stakeholder, but the IT rating does not. While the TYCOM responsibilities appear to work for the FC rating, they do not appear to work for the IT rating because of the multiple TYCOMs involved.

For the IT rating, while the NETC instruction identifies triggers, it is not clear whether any one of the triggers is strong enough to overcome the inertia resisting change. In contrast, there is sufficient interaction among AF cyber operations stakeholders to overcome organizational inertia. They talk constantly, and discussions are among the functional people, the trainers, and the major command user about making the specialty better to accommodate needed changes. The HM rating has similar interactions. Additionally, both of these communities have identified a single individual with ultimate responsibility for the improvement of these communities. In both cases, that individual is of sufficient rank to initiate change and to cause decisionmaking to occur.

In summary, the IT rating's MPT system does not respond to urgency or legitimacy. This is partly because the IT rating's performance is not viewed as an urgent issue within the Navy. IT personnel serve in relatively small numbers in many units, and thus IT personnel performance increases would result in only marginal performance increases for these units. As a result, there is insufficient urgency regarding IT performance. The PEO C4I has a stated interest in IT personnel performance due to the effect on C4I system performance, and the PEO has legitimacy in certain OPNAV instructions but lacks legitimacy in the manpower, personnel, and training environments where decisions are made. Further, in the case of the IT rating, the power (money) to influence change is spread across several resource sponsors and across users represented by multiple type commanders, each of whom has competing priorities for resources.

Recommendations

Navy leadership should be cognizant of the goal orientation and strategy of organizations in order to assess whether those goal orientations are the most appropriate for the organizations. Further, awareness of differing goal orientations can facilitate interactions by bringing explicit awareness of differing stakeholder strategies. Our recommendations are specific to the IT community but might also apply to other Navy communities.

The IT community should have a single identified stakeholder with responsibility for performance and MPT oversight. Ideally, this stakeholder would have sufficient legitimacy and power (money) to influence change. Since the IT community has many different users, this predominant stakeholder should interact with all users as part of an institutional process for gathering feedback from users and making improvements. Organizationally, who should the stakeholder be and how could this process be implemented? For the HM community, the BUMED organization has the legitimacy of OPNAV instruction authority, the power of the purse (resourcing of training), and urgency as the user of the outcome of training. This model has appeal, and early documentation for the creation of an "Information Dominance Corps" cited such a model, but it has not eventuated. The FC model appears currently workable for that community because of a strong system command, resource sponsor, and TYCOM, with stable, consistent relationships. The IT stakeholders, whether enterprise, TYCOM, or OPNAV staff, have changed structural relationships considerably in recent years. Absent radical change to an HM community model or a change of a type we did not consider,[1] an AF model might offer the best path to effectiveness and efficiency. The N2/N6 would be the dominant stakeholder in this model, with responsibility for development and training of IT personnel. As in the AF model, this authority would be supported by functional authorities within N2/N6 and a career field manager in an operational unit, with authorities designated in Navy instructions. The N2/N6 would have ultimate responsibility for effectiveness of the community. This arrangement would need to be supported by more workable processes than those that currently exist.

For example, there should be regularly scheduled, periodic reviews of IT rating sustainment and new equipment training. For this community, a periodicity of 18 to 24 months (rather than three years) may be appropriate because of the expected rate of change in technology. As in the AF cyber operations community, these reviews should be under the auspices of the functional lead and chaired by the functional career manager, with support provided by NETC and the schoolhouse. These reviews should include users and should have the oversight of the N2/N6, the predominant stakeholder for decisionmaking and resourcing. Such reviews and resulting periodic changes could move the IT MPT community toward effectiveness as an outcome.

At the TYCOM level, IT rating users should have the opportunity and capability to make efficiency and effectiveness decisions. For example, the opportunity to have fewer, better-trained personnel should be one such decision regularly considered. Because of the number

---

[1] For example, if N2/N6 were assigned as training resource provider and with TYCOM responsibilities, with SPAWAR as training provider. This could have merit to consider. In particular, because the IT community appears to have more technology change and thus more new equipment training than many other communities, it may increase effectiveness and efficiency to have one organization responsible for resourcing and for executing both sustainment and new equipment training.

and diffused nature of the user stakeholders, the functional authority (N2/N6) needs to take the prevalent role in these manpower reviews, supported by the N12 manpower community.

While these recommendations are consistent with good management practices, we recognize that it will be difficult to improve the effectiveness of IT personnel for several reasons. First, because they are generally diffused throughout the Navy in small numbers, there may not be users who will see significant organizational performance increases because of increased IT personnel performance. Second, IT personnel do not operate or maintain a Navy combat system. Unlike the AF, which has recognized and publicly stated that information technology is vital to its mission performance, the Navy has not prioritized IT personnel performance.[2] Given the institutional emphasis on efficiency foremost, additional investments in IT performance are currently unlikely without significant changes in organizational structure and processes and in stakeholder responsibilities.

---

[2] The Chief of Staff of the AF has stated that "cyber operations reinforce and enable everything we do, from administrative functions to combat operations, and we must treat our computers and networks similarly to our aircraft, satellites, and missiles" (*Cyberspace—Shaping the New Frontier*, 2010). The Navy considers its computing networks to be in support of Navy weapons systems.

APPENDIX A

MPT Stakeholders and Relationships

This appendix has four sections. The first is an introduction and an assessment of the evolution of Navy training responsibilities. Continual change in these relationships might indicate broader institutional issues with training that affect all Navy communities. This section is followed by a discussion of NTSPs in general and then for the CANES program in particular. Last is a discussion of sustainment training responsibilities.

Introduction

The Navy has a process and established policies and procedures for planning, determining, and documenting MPT requirements for new and modernized acquisition systems across the continuum of training (ashore, pier-side, and afloat). This process and these policies delineate responsibilities for cognizant offices. As the Navy has reorganized and shifted responsibilities, offices (or stakeholders) responsible have changed or evolved over time.

Joint Capability Integration and Development System documents guide the development of future acquisitions and establish required system, operator, and maintainer capabilities. These capabilities also include key performance parameters and establish system, operator, maintainer, and employment capabilities. OPNAV instruction 1500.76B[1] is used to develop training planning in support of new or modernized Navy capabilities. NTSPs are planning products developed during DoD acquisition processes.

PMs must perform a front-end analysis (FEA) to include a job task analysis and workload analysis for new and/or modified systems. The FEA process examines the new or modified system and determines what manpower and knowledge, skills, and abilities (KSAs) will be required to safely operate and maintain these systems and equipment. A preliminary NTSP is developed by the PM to document MPT requirements identified during the FEA.

Resource sponsors are required to obtain concurrence from N15 prior to approving a final NTSP. As programs change, NTSPs must be reviewed annually and updated throughout a system's life cycle, to include modernization.

[1] Department of the Navy, Chief of Naval Operations, 2010.

Evolution of Naval Training System Requirements, Acquisition, and Management

The Navy's original instruction (Department of the Navy, Chief of Naval Operations, 1998) detailed training roles and responsibilities for new and/or modified systems and the offices responsible for executing them. This instruction has been updated twice, in 2006 and 2010. The updated instructions shifted some responsibilities among organizations. Changes in Navy organization over time account for many of these shifts. The change in responsibilities brings different stakeholders into the process, with varying responsibilities, views, and approaches.

We examined Navy guidance to determine the similarities and differences with regard to the offices responsible over time. Table A.1 illustrates the evolution of the offices responsible from the original through the current instruction.

The similarities over time as reflected in the guidance are that N1, resource sponsors, program managers, training support agency (TSA), training agency (TA), and systems commands have maintained training responsibilities over time. However, N12, N43, N15, the Fleet Forces Command (FLTFORCOM), and enterprise commanders[2] have since 2006 been assigned responsibility for Navy training system requirements.

Some offices that have had a past role in naval training system requirements no longer have responsibilities in the process, including Chief of Naval Operations N7, the Director of the Naval Reserve, the training effectiveness evaluation agent, the Training Planning Process Methodology Advisory Board, fleet project teams, the Deputy Assistant Secretary of the Navy for Civilian Personnel/Equal Employment Opportunity, the Commandant of the Marine Corps, the Marine Corps Representative, Fleet Commanders in Chief, and the Naval Safety Center. The TYCOM's role has shifted to that under its Surface Warfare Enterprise role, noted above.

The evolution of the roles can be attributed to Navy reorganization—combining N7/N8 functions, the increased role of OPNAV N15 in sustainment training, and decreased roles for Fleet Commanders in Chief and for other responsible offices due to realignment of responsibilities. The Surface Warfare TYCOM executes his MPT role for new/modernized systems as the Surface Warfare Enterprise Commander.

The process has evolved to a reduced number of offices responsible for naval training system requirements.

Navy Training System Plans

NTSPs are documents that communicate MPT gaps and needs in support of new acquisition and/or modernization programs. Officer, enlisted, and civilian manpower and qualification requirements are documented in NTSPs.

NTSPs ensure adequate planning, programming, and budgeting of sustainment training throughout the future years defense program. Resource sponsors must get concurrence from OPNAV N15 prior to approving a final NTSP. Once the resource sponsor approves a final or

[2] The Surface Warfare Enterprise Commander is the Surface TYCOM, which is Commander, Surface Forces, Pacific.

Table A.1
Offices Responsible for Naval Training System Requirements, Acquisition, and Management in OPNAVINST 1500.76 Series Instructions

Office	OPNAVINST 1500.76 (1998)	OPNAVINST 1500.76A (2006)	OPNAVINST 1500.76B (2010)
N1	√	√	√
N12		√	√
N43		√	√
Chief of Naval Operations N7	√		
Director of the Naval Reserve	√		
N15			√
Program sponsor/resource sponsor	√	√	√
Program manager	√	√	√
Training support agency	√	√	√
Training agency	√	√	√
NETC		√	√
Naval Systems Command	√	√	√
PEO	√	√	
Direct reporting program manager	√	√	
Developing Activity	√	√	
Training effectiveness evaluation agent	√		
Training Planning Process Methodology Advisory Board	√		
Fleet project teams	√		
Deputy Assistant Secretary of the Navy for Civilian Personnel/Equal Employment Opportunity	√		
Commandant of the Marine Corps	√		
FLTFORCOM		√	√
Marine Corps Combat Development Command	√	√	
Marine Corps Representative		√	
Fleet commanders in chief	√		
TYCOMs		√	
Naval Safety Center	√		
Enterprise commanders			√

updated NTSP, the NTSP shall be used as the official record of the training planning process that facilitated the enterprise definition of the system's MPT requirements.[3]

There are timelines and deadlines associated with the NTSP development and approval process, and responsibilities are assigned. PMs must perform the FEA—which is the process to determine gaps between required and existing knowledge, skill, and ability requirements for all dissimilar, new, or modified equipment/system/subsystem; to compare these requirements to existing training; to define acquisition and development requirements; and to develop a strategy to most cost-effectively fill any identified gap.[4]

Navy activities are assigned roles and responsibilities in the NTSP development process and are broadly described in Table A.2. OPNAV N1 organization (N1, N12, and N15) are key players as they assess the impact of a new system/program on N1 resources (N1), verify manpower estimates and KSAs for new or modified systems (N12), and act as resource sponsor for sustainment training (N15). OPNAV N43 resources unit-level training. Resource sponsors provide funding for the updated training system and all MPT requirements until the TA takes over. System commands give authority to the PMs to pursue development of new or modernized systems. TSAs support curriculum development to meet the training requirements of the systems. TAs execute the training, and NETC identifies training infrastructure and facilities and the instructors needed to address training needs. FLTFORCOM validates, prioritizes, and provides oversight and resources for fleet training. Enterprise commanders work with FLTFORCOM, resource sponsors, and the TA to identify training requirements necessary to meet training conditions and standards.

Organizations That Are NTSP CANES Principals

SPAWAR, PMW 120/150/160 are the Training Support Agencies (TSAs). SPAWAR is responsible for the Lifecycle Maintenance Plan and PMW 160 provides instructor (uniformed) manning or Subject Matter Experts (SMEs) if needed. If those aren't suitable, PMW 160 can go to a third party vendor with TA approval. They develop end to end training requirements and evaluate end to end training proficiency. (Department of the Navy, Space and Naval Warfare Systems Command, 2009) Initial training, often associated with installation training, is the system specific training developed and delivered by the TSA in accordance with a fleet stakeholder validated solution prior to the training transition to the TA (NETC/CID). Additionally, as new technologies not covered in NETC networking foundation training courses are introduced with new upgrades to the CANES system, the TSA delivers new technology core familiarization training to supplement initial training. PMW (Program Manager Warfare) 160 develops the CANES PQS, covers logistics support requirements for manufacturers, develops technical manuals, writes Human Systems Integration Plan, and does onsite installation training for all new COTS equipment. The SPAWAR Institute, now the TDSC, supports the PEO PMs in meeting their TSA responsibilities.[5] (Department of the Navy, Space and Naval Warfare Systems Command, 2009) Overall, the Program Manager writes objectives for develop-

[3] Department of the Navy, Chief of Naval Operations, 2010.

[4] Department of the Navy, Chief of Naval Operations, 2010, p. 6.

[5] Since this research, the SPAWAR Institute has been changed to the SPAWAR Systems Center, Pacific (SSC PAC), C4ISR Training Development and Support Center (TDSC).

Table A.2
Roles and Responsibilities of Activities Involved in MPT for New or Modernized Navy Systems

Activity	Role and/or Responsibility
N1 (Manpower, Personnel, Training and Education)	Validates impact of new program on total N1 resources; plan/budget for MPT requirements; serves as manpower resource sponsor
Director, Total Force Requirements Division (OPNAV N12)	Validates manpower documents and coordinates programming and budgeting issues with enterprise; approves modified manpower estimates; validates occupational standards needed by the system commanders of KSAs needed for officers and enlisted personnel
OPNAV N15	Validates individual training as resource sponsor for training sustainment; ensures NTSP requirements support mission requirements; ensures that coordination of interim plans for initial training is addressed
Director, Fleet Readiness Division (OPNAV N43)	Provides fleet readiness training resources and oversight
Resource sponsors	Plan, program, fund requirements for design, development, procurement, engineering change, and modernization of the training system for the life cycle; fund all training until the training system is implemented and accepted by TA; assess MPT supportability in coordination with OPNAV N15; fund all MPT requirements until transfer of responsibility to N1
Systems commands	Empower PMs with authority to acquire, modernize, and maintain configuration management of the total training system
Program managers	Identify, plan, budget, and submit all system and resource requirements, including NTSPs; perform FEA; identify and implement approved training resource requirements; program and budget resources to provide initial or other training identified in NTSP
TSAs	Develop/implement training system installation plans to meet ready for training date (RFT); develop curriculum, courseware and content
TAs	Execute approved training system sustainment requirements at RFT date; execute follow-on training requirements of training system; validate training curricula; plan, budget, and initiate POM and PR submission to the acquisition program resource sponsor
NETC	Plans and executes TA responsibilities when designated; identifies requirements for training infrastructure and instructors; validates FEA identification of MPT gaps; communicates effectiveness of a training solution to satisfy capability requirements
USFFC	Validates and prioritizes the fleet training program; provides policy, guidance, oversight, and programmatic resources for fleet training
Enterprise commanders	Provide USFFC, resource sponsors, and TA with operational requirements, feedback, and lessons learned on effectiveness of a training solution to satisfy capability requirements; provide guidance to PM on training requirements; identify training gaps and requirements; identify conditions or standards that meet training key performance parameters and/or KSAs for development of training

ment, production, and sustainment to meet the user's operations needs, and identifies training system shore facilities' requirements (including planning and installation). (Department of the Navy, Chief of Naval Operations, 2006)

PEO-C4I is the Developing Activity (DA) and coordinates the TSAs for the training modules and curriculums across the system's TSAs. (Department of the Navy, Space and Naval Warfare Systems Command, 2009) It establishes the TRPPM board. (NTSP CANES) According to OPNAV Instruction 1500.76A, the PEO (Developing Activity), Direct Reporting Program Manager, undertakes the management and technical responsibility for assigned acquisition programs. It identifies and plans all system requirements, including the NTSP, manpower documents, and ensures inclusion of MPT resource requirements in planning, programming, budgeting, and execution (PPBE). PEO develops on board (in-service) training, programs and budgets resources for curricula and training materials, provides technical manuals and documentation, provides the TA with all manuals, curricula, maintenance requirements cards, assist modules, initial repair parts, and technical assistance for CASREPs, and funds and procures diagnostic software for training equip. (Department of the Navy, Chief of Naval Operations, 2006)

SYSCOM, PEO, PMs are usually the Development Activity (PEO-C4I for CANES), provide all training support, and develop and implement life cycle support for training systems. They work with fleet for onboard training and interoperability and budget resources for initial curricula. (Department of the Navy, Chief of Naval Operations, 1999) For interoperability of C4ISR systems, they ensure joint information standards comply and are in the design of C4ISR systems, request waivers from N6 when current systems cannot interface with new systems, put lifecycle fiscal planning of interoperability in program budgets, and coordinate with the Navy Center for Tactical System Interoperability (NCTSI) for identifying interoperability requirements. Program Managers (PMs) must apply for an interim certificate to Navy Interoperability Test Panel (ITP) representative for delays and get recertified with software upgrades every 3 years from NTCSI. (Department of the Navy, Deputy Chief of Naval Operations (Communication Networks) (N6), 2008)

N61 is the OPNAV Resource Sponsor and approves manpower and training resource requirements and the CANES NTSP. N61 also serves as the OPNAV Principle Official (OPO) and **N612** (Command and Control Systems Division) is the Functional Mission Sponsor. All three positions are NTSP Principals. The OPNAV Resource Sponsor funds development and delivery of NETC (TA) Foundation Training updated for the system before the IT of the Future training update projected in 2010. OPNAV N6 and the TSA fund and provide new technology core training with fleet introduction of a new system/variant before introduction to the TA. (Department of the Navy, Space and Naval Warfare Systems Command, 2009) N6 submits interoperability standards recommendations to JCS, NSA, and DISA, ensures system compliance with joint interoperability standards, and funds NCTSI configuration management, test support, fleet support, and network design facility functional requirements. (Department of the Navy, Deputy Chief of Naval Operations (Communication Networks) (N6), 2008) OPNAV Resource Sponsors plan, program, and fund resources to transition legacy systems to joint/navy information standards during upgrades. They also address interoperability requirements in requirements documents and fund testing (Department of the Navy, Deputy Chief of Naval Operations (Communication Networks) (N6), 2008)

N1 (Deputy CNO (Manpower Personnel, Training, and Education)) is the Director of Naval Training, the MPT Mission Sponsor, and therefore an NTSP Principal. (NTSP CANES) N1 coordinates with N2/N6 to manage the information community (and their programs). It is the only manpower resource sponsor that manages funding and resources for total force manpower and personnel, navy requirements, and authorization. (Department of the Navy, Chief of Naval Operations, 2009) N1 evaluates the impact of acquisition and modernization programs on total Navy MPT. This evaluation makes sure that costs, programming and collaboration with NTSP principals (N61, N612, PEO-C4I, NETC, CID, SPAWAR, N1) agree with long-term MPT plans. It receives a Manpower Estimation Report from the SPAWAR Institute. N1 is also notified by the Program Managers (TSAs) of student throughput requirements early in the program planning stage (Department of the Navy, Chief of Naval Operations, 2006). N15 executes N1 NTSP responsibilities for training transition approval. Also, see N12 section below.

Navy Education and Training Command (NETC) and the Center for Information Dominance (CID) are the Training Agencies (TAs). They serve together as a NTSP Principal. NETC provides CANES foundation training, Technical Training Equipment (TTE), and Information Technology Computer Based Training courses on-line or via local network installation. (Department of the Navy, Space and Naval Warfare Systems Command, 2009) They provide and resource their staff for initial pilot and product evaluations, and common core instructor training. They are responsible for staffing annual course reviews and courseware updates and are required to conduct returns on investment for longer term training. NETC and CID approve Training Support Agency (SPAWAR) designed training for new complex updates to systems, and determine whether a TSA can use Subject Matter Experts (SMEs) or if they need a third party vendor. (Department of the Navy, Space and Naval Warfare Systems Command, 2009)

Additionally, the TA shall do NTSP development, review training curricula, transfer responsibility of support equipment after delivery, and implement lessons learned on training effectiveness. NETC plans and executes responsibilities assigned to the TA when designated (NETC, Naval Personnel Development Command (NPDC), and Learning Centers), funds data item descriptions and contract data requirement lists, coordinates learning center efforts, and conducts NTSP reviews. NPDC does maintenance and contractor operations for simulators. (Department of the Navy, Chief of Naval Operations, 2006)

Other Organizations That Are Not Directly NTSP CANES Principals

N12 is also known as the Total Force Programming, Manpower Management Division and fulfills the role of principal program and manpower sponsor. One of its primary responsibilities is to approve manpower estimate requirements. N12 assists the Program Manager (PM) with the Manpower Estimation Report, maintains training fiscal costs estimates every year on file, and provides oversight as a member of the advisory boards for NTSP policy. Additionally, N12 provides strategic direction for individual skills and human performance, assesses training programs for human objectives, ensures compliance with joint forces, and monitors fleet training readiness through the Naval Personnel Development Command (NPDC) Personal Development Advisory Group (PDAG) process. (Department of the Navy, Chief of Naval Operations, 2006)

N43 is known as the Fleet Readiness division and provides fleet training oversight. It approves NTSPs after N12 concurrence, programs funding for approved training requirements, and programs reviews to make sure MPT is met. Also, N43 evaluates the fleet commanders, type commanders, and subordinate commanders' feedback and lessons learned on NTSP implementation. (Department of the Navy, Chief of Naval Operations, 2006)

Type Commanders (previously Naval Network Warfare Command (NAVNETWARCOM) and now Navy Cyber Forces Command in this case) serve on the Training Planning Process Methodology (TRPPM) Advisory Board and is a Naval Training System Plan Conference (NTSPC) member. They advise program sponsors, resource sponsors, N12, Program Managers, and the Naval Personnel Development Command (NPDC) when NTSPs are no longer synchronized. They provide guidance to the Program Manager on training requirements and participate in everything on weapon systems (life cycle, design, introduction, etc.).

They provide Fleet Forces Command (FFC) Program Sponsors, resource sponsors, and the Training Agency (TA) with operational requirements, feedback and lessons learned on the effectiveness of training solution to satisfy the capability requirement. (Department of the Navy, Chief of Naval Operations, 2006)

SHIP is responsible for on-board training that will be provided by Interactive Courseware (ICW) and Integrated Electronic Technical Manual (IETM) for PQS at the Unit Level (ship). (Department of the Navy, Space and Naval Warfare Systems Command, 2009)

N2/N6 (Information Dominance Corps) advises CNO, US Fleet Forces Command, CNP regarding issues associated with N2/N6. N2/N6 also provides recommendations to N1 on all cultural and community issues that affect MPT, serves as the subspecialty system major area sponsor for all curricula in NAVPERS 15839I, and creates a quarterly flag panel to collaborate across all communities within N2/N6 for slating, policy, and programmatic issues. (Department of the Navy, Chief of Naval Operations, 2009)

CNP (Chief of Naval Personnel) sets personnel policy for community management, recruitment, distribution, and advancement. Deputy CNP does analysis and manpower forecasting, assigns sailors to job vacancies, and sponsors selection boards. (Department of the Navy, Chief of Naval Operations, 2009)

N86 (Director, Surface Warfare Division) is the program and resource sponsor for training systems, reviews all systems to ensure MPT requirements are adequately defined, provides oversight to acquisition program for MPT issues, and funds the three previous listed items. N86 is responsible for the development, approval, and life cycle management of NTSPs. N86 coordinates activities with N1 (ensures manpower is fully funded and 100% manning), N4 ((Deputy CNO Logistics) ensures infrastructure investments decisions that guarantee interoperability and supportability for new and legacy systems), N6 ((Director, Space and Information Warfare) coordinates development of C4I systems, distance support learning and training), N7 ((Director, Naval Training) improves training ashore and support for On-the Job Training (OJT) onboard), and N8 ((all N8* warfare codes) shares MPT initiatives to support effective resource utilization) (Department of the Navy, Chief of Naval Operations, 1999)

Fleet Commanders communicate requirements to the resource sponsor (N61) and program manager (PMW 160), provide oversight of training functions, and act as final customer to identify inadequacies. (Department of the Navy, Chief of Naval Operations, 1999) Fleet Commanders ensure C4ISR systems incorporate approved interface and data standards (ones that are tested from NCTSI or certified by JITC), and notify N6 and NCTSI when a system is rapid prototyping or in the fleet initiative program as compared to the standard process. (Department of the Navy, Deputy Chief of Naval Operations (Communication Networks) (N6), 2008)

NCTSI (Navy Center for Tactical System Interoperability) is the primary navy activity for technical support to the CNO for configuration control and interoperability. It coordinates with fleet and system commands for C4ISR interfaces (e.g., link 16), is a voting member on IPT and other joint committees, develops a navy interoperability configuration management plan for procedural interface standards, establishes criteria to certify C4ISR program compliance with interface standards, and conducts interoperability testing as requested by the PM. Additionally, NCTSI does CERT testing of interoperability which it coordinates with the PM and sends to COMNETWARCOM and CNO, coordinates with SYSCOMs and PEOs to do parallel testing, and serves as Software Support Activity (SSA) and In-Service Engineering

Agent (ISEA) for the Navy program of record test tool for C4ISR systems. (Department of the Navy, Deputy Chief of Naval Operations (Communication Networks) (N6), 2008)

N8 (Deputy CNO, Integration of Capabilities and Resources) submits a requirements document for C4ISR systems to the joint staff for interoperability certification. (Department of the Navy, Deputy Chief of Naval Operations (Communication Networks) (N6), 2008)

COMOPTEVFOR (Commander, Operational Test and Evaluation Force) cites operational test results in an evaluation report to CNO, COMNAVNETWARCOM, PEO C4I PM, SPAWAR, and NTCSI. COMOPTEVFOR also attends Navy program decision meetings and ensure technical standards testing interoperability certification testing happen before Operational Evaluation (OPEVAL) and Follow-On Test and Evaluation (FOT&E). (Department of the Navy, Deputy Chief of Naval Operations (Communication Networks) (N6), 2008)

Roles and Responsibilities for Navy Sustainment Training

The following are the Navy activities and/or organizations with sustainment training responsibilities for new and/or modernized systems. Training responsibilities were derived from the most current Navy guidance—Department of the Navy, OPNAV Instruction 1500.76B, Naval Training System Requirements, Acquisition, and Management, 28 April 2010.

Activity	Role/Responsibility
OPNAV N1	Plan, program, and budget for sustainment of MPT requirements following ready for training (RFT) date.
OPNAV N15	Validate individual training requirements as the resource sponsor for sustainment of individual training requirements after RFT date. Participate in various SYSCOM logistics, training support agency (TSA), United States Fleet Forces Command (USFLTFORCOM) (Director, Manpower and Personnel (NI) and Director, Training (N7, Naval Education and Training Command (NETC) (Director, Navy Training (N7, and enterprise conferences to resolve MPT acquisition and life cycle issues.
OPNAV N43	Provide training resources and oversight.
Program Managers (PM)	Submit funding requirements to the OPNAV (N15) and Training Agency for Tactical Training Equipment (TTE), Training Devices (TD), and simulator/stimulator COMS requirements (a minimum 2 years prior to RFT). Develop training for major revisions required due to engineering change proposals and/or modifications to system(s). As program changes dictate, and at a minimum annually, PMs review NTSPs to determine if an update is justified and report their findings to the resource sponsor. Coordinate with the TA responsible for follow-on training. Arrange interservice training support, if required. Program and budget for alteration, conversion, and restoration of TA training facilities when installing and removing training equipment. Program and budget to develop, procure, deliver, install, overhaul, and modernize TTE, TD, stimulators and other training material requirements identified in the NTSP throughout the life cycle of the system. Develop technical manuals, documentation, and updates for use in initial and follow-on training. Notify the resource sponsor, OPNAV (N15), USFLTFORCOM, the fleet user (i.e., fleet commander), the TSA and the TA, in sufficient time to allow appropriate risk mitigation action (e.g., manpower, equipment, and resources) in the event that a training solution *is* not adequately funded. Participate in training effectiveness evaluations (TEEs) as requested by the resource sponsor.

Activity	Role/Responsibility
Resource sponsors	Prior to approving a final or updated NTSP, obtain concurrence from OPNAV (N15) to ensure adequate planning, programming, and budgeting of sustainment training throughout the FYDP. Assess MPT supportability of all acquisitions and modernizations and provide resolution in coordination with OPNAV (N15) for all MPT issues. Identify changes impacting MPT that result from programming, reprogramming, budget changes, development or production schedule changes, equipment modernizations, manpower and personnel, life cycle maintenance, and management changes impacting training for the life cycle of the weapon or supporting system. Notify OPNAV (NI5) of any changes impacting MPT. Initiate appropriate action to correct MPT issues identified by USFLTFORCOM, fleet commanders, warfighter enterprises, and subordinate commander staffs. Fund the training requirements prior to transition of individual training resource sponsorship at RFT, and after for engineering change and modernization of the training solution for the life cycle of the system(s). Oversee the planning, development, implementation and effectiveness of the training system installation plan (TSIP), leading to the TA final acceptance of the training system from the TSA to meet the established RFT date. Maintain formal liaison with OPNAV (N15, USFLTFORCOM, fleet commanders, Commander, Naval Reserve Force, applicable TSA/TAs and inter-service agencies or components to achieve satisfactory final acceptance of the training system at RFT. Report the results of annual NTSP reviews to OPNAV (N15) by 1 February of each year.
Training Support Agency (TSA)	Develop training solutions that equip a Sailor with the proper knowledge, skills, and abilities to meet fleet/Fleet Marine Force (FMF) requirements. Develop curriculum, courseware and content, including distributed learning, in-service training, onboard training, and self-paced computer-based training. Conduct HSI planning, FEA, and develop required PQS for associated MPT requirements. Coordinate with the TA responsible for follow-on training to determine an RFT date recommendation to the resource sponsor for approval and programming.
Training Agency (TA)	Execute approved training system sustainment requirements at RFT date. Submit all Tactical Training Equipment (TTE), Training Device (TD), and simulators/stimulators sustaining resource requirements to Training Support Agency (TSA). Execute follow-on training requirements of the approved training system. Execute responsibilities of PQS model manager after fleet introduction. Provide SMEs for FEA, MPT advisory boards, and NTSPCs. Validate training curricula and materials to ensure initial training and follow-on training meets requirements. Perform all assigned steps and responsibilities to participate in the planning, development, implementation, and review of the TSIP, leading to TA final acceptance of the training system from the TSA to meet the resource sponsor established RFT date. Execute the transition of responsibility for the training solution(s), including TTE, TDs, simulators/stimulators, and support equipment after acceptance testing and delivery. Submit to TSA all TTE, TD, and simulators/stimulators sustaining resource requirements. Submit to the PM, USFLTFORCOM, fleet commander, and subordinate commander staff feedback and lessons learned on training effectiveness.
NETC	Validate that the program's FEA identifies the gaps between baseline comparison MPT requirements and new equipment/system/sub-system MPT requirements. Develop a comprehensive training system acquisition strategy. Approve TDs/simulators/stimulators consistent with that strategy, and provides military characteristics documents for these trainers to support timely POM submissions. Make recommendations to OPNAV (N15) for future POM submissions for post RFT sustainment requirements. Validate that training requirements listed in NTSPs are included in officer and enlisted school and skill development plans. Coordinate and resource subordinate activities and learning centers for participation supporting the NTSP process. Standardize, integrate, and support individual training and education of the warfighter. Coordinate with respective enterprise to determine the adequacy of individual skills training as it impacts mission capability and fleet/FMF readiness. Develop, update and modify individual training to meet fleet requirements. Communicate with USFLTFORCOM, warfighting enterprises, fleet commanders, subordinate commander staffs, resource sponsors, and the PM for feedback and lessons learned on the effectiveness of the training solution to satisfy the capability requirement.

Activity	Role/Responsibility
USFLTFORCOM	Validate and prioritize fleet training program. Provide policy, guidance, oversight, and programmatic resources for fleet training. Provide guidance to subordinate activities to participate in the NTSP development and review process.
Enterprise commanders	Provide USFLTFORCOM, resource sponsors, and the TA with operational requirements, feedback, and lessons learned on the effectiveness of the training solution to satisfy the capability requirement. Identify individual, team, and fleet operator and maintainer training gaps and requirements, and report findings as required to resource sponsor, OPNAV (N151), PM, and appropriate learning center.
Type commander	Determine/establish and list individual training requirements. Provide units with resourcing for individual training requirements.
Unit	Schedule, coordinate and execute individual sustainment training requirements.

APPENDIX B

Methodology for Effectiveness and Efficiency Goal Orientations

This appendix provides additional detail regarding the methodology used to calculate an effectiveness and efficiency goal orientation for the IT stakeholders. We drew 38 statements from the literature (largely, Burton, DeSanctis, and Obel, 2006) that could be used to assess whether a stakeholder had an effectiveness or efficiency orientation. Each statement was then associated with a quadrant (or quadrants) of the strategy framework shown earlier in Figure 4.1. For example, the upper right quadrant in Figure 4.1 would have positive values for x and y while the lower left would have negative values for x and y.

A five-point scale, ranging from completely disagree (1) to completely agree (5), was used. Table B.1 shows the quadrant placement for each possible scale score, 1 to 5, for each of the 38 items. This table indicates the x,y value for each possible answer on the five-point scale for each statement. In each instance, the first column indicates the x value and the second column indicates the y value. For example, the second statement in the table is "Can make large changes in products and services." Organizations that can clearly do this would be assigned a score of 5, or strongly agree. Those organizations would be given a 1, –1 (shown at the right of the table) that would reflect an "experiments" strategy. Those organizations that received a score of 1, or strongly disagree, would receive a –1, 1. These statements result in both an x and a y score because the statement is attributed to a particular strategy, or quadrant.

As another example, the first statement in the table is "Little focus on using resources well." This statement is one that pertains to the efficiency axis of Figure 4.1, rather than to a single strategy. Thus, organizations that were scored a 5 for this statement were given a –1 for the y-axis but were not located on the x-axis. Likewise, organizations that received a 1 for this statement were scored a 1 on the y-axis.

The last four statements assess the environment of the organization. These four statements are mutually exclusive such that the researchers cannot envision an organization that could receive a positive answer (4 or 5) on more than one of these questions. Instead, most organizations will receive a 5 on one of the environment statements and a 1 on the remaining three environment statements.

For this analysis, the researchers used a linear score, such that a score of 2 was weighted half as much as a score of 1, a score of 3 received zeros and a score of 4 was weighted half as much as a score of 5. This analysis explored the sensitivity of nonlinear weighting schemes, such that a score of 2 would equate to 40 percent of a score of 1, or essentially either 0.4 or –0.4. Doing so changed the answer minimally, and the researchers did not have a valid reason to weight the evaluations nonlinearly.

Given the values in Table 4.1, each statement was then assessed for each organization by the researchers using the five-point scale. These assessments were based on interviews with individuals in the organizations, observed processes and policies, and expert knowledge of these organizations. These assessments were subjective, and others may disagree with some of the evaluation. As we acknowledge that our scoring is subjective but based on interviews, reports, instructions, and observations, we provide the methodology that we used, rather than our actual scoring, so that readers can perform their own assessment if desired.

After developing the score calculations for each statement for each organization, the researchers averaged the scores and plotted the organization's goal strategy with a scale of –1 to 1 on both the x-axis (effectiveness) and the y-axis (efficiency). The results of this plot for the organizations we assessed are shown earlier in Figure 4.2.

Table B.1
Strategy Evaluation Statements and Score Calculations

	Scale									
	Quadrant Placement 1		Quadrant Placement 2		Quadrant Placement 3		Quadrant Placement 4		Quadrant Placement 5	
Statement	x	y	x	y	x	y	x	y	x	y
Little focus on using resources well		1		0.5		0		–0.5		–1
Can make large changes in products and services	–1	1	–0.5	0.5	0	0	0.5	–0.5	1	–1
Lacks an intentional strategy toward innovation	1	1	0.5	0.5	0	0	–0.5	–0.5	–1	–1
Creator of change; other organizations must adjust to its actions	–1	1	–0.5	0.5	0	0	0.5	–0.5	1	–1
Has a reputation as an innovation leader	–1	1	–0.5	0.5	0	0	0.5	–0.5	1	–1
Adjusts processes when ordered or when there is an urgent need or problem	1	1	0.5	0.5	0	0	–0.5	–0.5	–1	–1
Adjusts to changed situation after the fact; reacts	1	1	0.5	0.5	0	0	–0.5	–0.5	–1	–1
Makes decisions based on "bad news" as it becomes known	1	1	0.5	0.5	0	0	–0.5	–0.5	–1	–1
Few or no specific goals related to broad objectives	1		0.5		0		–0.5		–1	
The new and different are more important than quality or price	–1	1	–0.5	0.5	0	0	0.5	–0.5	1	–1
Does not systematically anticipate, plan, or look to the future	1	1	0.5	0.5	0	0	–0.5	–0.5	–1	–1
Has an active innovation strategy for developing new products, services, and processes	–1	–1	–0.5	–0.5	0	0	0.5	0.5	1	1

Table B.1—Continued

	Scale									
	Quadrant Placement 1		Quadrant Placement 2		Quadrant Placement 3		Quadrant Placement 4		Quadrant Placement 5	
Statement	x	y	x	y	x	y	x	y	x	y
Problems emerge as surprises and are dealt with as they occur	1	1	0.5	0.5	0	0	–0.5	–0.5	–1	–1
Primary focus is using the smallest amount of resources needed to produce products and services	1	–1	0.5	–0.5	0	0	–0.5	0.5	–1	1
Experiments with new internal processes	–1	1	–0.5	0.5	0	0	0.5	–0.5	1	–1
Focus on process innovation with efficiency as primary goal	1	–1	0.5	–0.5	0	0	–0.5	0.5	–1	1
Focuses on preserving consistency and maintaining reputation	1	–1	0.5	–0.5	0	0	–0.5	0.5	–1	1
Cannot change much or change quickly; slow to make change	1	–1	0.5	–0.5	0	0	–0.5	0.5	–1	1
Continues to do what it has done in the past while refining for continued improvement (i.e., no leaps ahead, not much progress)	1	–1	0.5	–0.5	0	0	–0.5	0.5	–1	1
Has a copy strategy; important to recognize what others are doing successfully	–1	–1	–0.5	–0.5	0	0	0.5	0.5	1	1
Focuses on strategic goals rather than efficient use of resources	–1	1	–0.5	0.5	0	0	0.5	–0.5	1	–1
Constantly develops new ideas or concepts; treats cost as a secondary concern	–1	1	–0.5	0.5	0	0	0.5	–0.5	1	–1
Examines customer needs and tries to innovate to meet those needs	–1	–1	–0.5	–0.5	0	0	0.5	0.5	1	1
Constantly questions the status quo	–1	1	–0.5	0.5	0	0	0.5	–0.5	1	–1
Exploits only current resources and situation; little exploration into new products or services	1	–1	0.5	–0.5	0	0	–0.5	0.5	–1	1
Experiments regularly with new ideas, new technology, or new processes	–1	1	–0.5	0.5	0	0	0.5	–0.5	1	–1
Adept at doing much the same thing efficiently with a few significant changes from time to time	–1	–1	–0.5	–0.5	0	0	0.5	0.5	1	1

Table B.1—Continued

	Scale									
	Quadrant Placement 1		Quadrant Placement 2		Quadrant Placement 3		Quadrant Placement 4		Quadrant Placement 5	
Statement	x	y	x	y	x	y	x	y	x	y
Extrapolates the past as a basis for future planning	1	–1	0.5	–0.5	0	0	–0.5	0.5	–1	1
Focus is on internal organization and bureaucratic struggles (as opposed to product)	1	1	0.5	0.5	0	0	–0.5	–0.5	–1	–1
Develops new products, services, and their delivery processes	–1	–1	–0.5	–0.5	0	0	0.5	0.5	1	1
Purposely innovates and searches for new products and services	–1	–1	–0.5	–0.5	0	0	0.5	0.5	1	1
Makes changes in existing processes, products, or services to stay efficient	–1	–1	–0.5	–0.5	0	0	0.5	0.5	1	1
Surveys widely in technology and with customers to look for opportunities to improve	–1	–1	–0.5	–0.5	0	0	0.5	0.5	1	1
Invests in technology and capitalizes on results	–1	–1	–0.5	–0.5	0	0	0.5	0.5	1	1
Operates in a complex (many interdependent factors) but predictable environment	1	–1	0.5	–0.5	0	0	–0.5	0.5	–1	1
Operates in a simple (few factors) but unpredictable environment	–1	1	–0.5	0.5	0	0	0.5	–0.5	1	–1
Operates in a complex (many interdependent factors), volatile, unpredictable environment	–1	–1	–0.5	–0.5	0	0	0.5	0.5	1	1
Operates in a simple (few factors), stable, and predictable environment	1	1	0.5	0.5	0	0	–0.5	–0.5	–1	–1

Bibliography

Agle, B. R., R. K. Mitchell, and A. J. Sonnenfeld. (1999). "Who Matters to CEOs? An Investigation of Stakeholder Attributes and Salience, Corporate Performance, and CEO Values," *Academy of Management Journal*. Vol. 42, pp. 507–525.

Amburgey, T. L., D. Kelly, and W. P. Barnett. (1993). "Resetting the Clock: The Dynamics of Organizational Change and Failure," *Administrative Science Quarterly*. Vol. 38, pp. 51–73.

Austin, J. R., and J. M. Bartunek. (2003). "Theories and Practices of Organizational Development," in W. C. Borman, D. R. Ilgen, and R. J. Klimoski, eds., *Handbook of Psychology*, Volume 12: *Industrial and Organizational Psychology*. New York: Wiley. pp. 309–332.

Barnett, W. P., and G. R. Carroll. (1995). "Modeling Internal Organizational Change," *Annual Review of Sociology*. Vol. 21, pp. 217–236.

Bowen, D. E., and C. Ostroff. (2004). "Understanding HRM-Firm Performance Linkages: The Role of the 'Strength' of the HRM System," *Academy of Management Review*. Vol. 9, pp. 204–221.

Burton, Richard M., Gerardine DeSanctis, and Borge Obel. (2006). *Organizational Design: A Step-by-Step Approach*. Cambridge, UK: Cambridge University Press.

Carroll, G. R., and M. T. Hannan. (2000). *The Demography of Corporations and Industries*. Princeton University Press.

Cyberspace—Shaping the New Frontier. (2010). video. Washington, DC: AFTV. May 26. As of January 27, 2011: http://www.safxc.af.mil

Dent, E. B., and S. G. Goldberg. (1999). "Challenging Resistance to Change," *Journal of Applied Behavioral Science*. Vol. 35, pp. 25–41.

Department of Defense (DoD). (2011a). DoD News Release No. 010-11, January 6.

Department of Defense (DoD). (2011b). DoD News Release No. 013-11, January 6.

Department of the Navy, Chief of Naval Operations. (1989). OPNAV Instruction 1500.51B, *Total Force Training Strategy*. March 31.

Department of the Navy, Chief of Naval Operations. (1995). OPNAV Instruction 1500.69A, *Navy Training Requirements Review (NTRR)*. October 10.

Department of the Navy, Chief of Naval Operations. (1998). OPNAV Instruction 1500.76, *Naval Training Systems Requirements, Acquisition, and Management*.

Department of the Navy, Chief of Naval Operations. (1999). OPNAV Instruction 150.57A, *Surface Warfare Training Strategy*.

Department of the Navy, Chief of Naval Operations. (2006). OPNAV Instruction 1500.76A.

Department of the Navy, Chief of Naval Operations. (2009). OPNAV Instruction 5300.12, *The Information Dominance Corps*.

Department of the Navy, Chief of Naval Operations. (2010). OPNAV Instruction 1500.76B, *Naval Training Systems Requirements, Acquisition and Management*.

Department of the Navy, Commander, Naval Education and Training Command. (2010a). NETC Instruction 1500.8, *Human Performance Requirements Review (HPRR)*.

Department of the Navy, Commander, Naval Education and Training Command. (2010b). NETC Instruction 1510.4, *Job Duty Analysis Policy.*

Department of the Navy, Deputy Chief of Naval Operations (Communication Networks) (N6). (2008). OPNAV Instruction 941.5B, *Navy Interoperability Requirements, Certification and Testing of Navy Tactical C4ISR Systems Implementing Procedural Interface Standards.*

Department of the Navy, Naval Sea Systems Command. (2010). *Proposed AEGIS Weapon System Navy Training System Plan.* N86-NTSP-S-30-8512I.

Department of the Navy, Navy Personnel Command. (2011a). *Enlisted: Community Managers: Medical: Hospital Corpsman (HM),* "General Description." Web page. April 6. As of September 21, 2011: http://www.public.navy.mil/bupers-npc/enlisted/community/crypto_it/Pages/IT.aspx

Department of the Navy, Navy Personnel Command. (2011b). *Enlisted: Community Managers: Information Dominance: IT: Information Systems Technician (IT),* "General Description." Web page. August 15. As of September 21, 2011:
http://www.public.navy.mil/bupers-npc/enlisted/community/medical/Pages/default.aspx

Department of the Navy, Space and Naval Warfare Systems Command. (2009). *Naval Training System Plan for Consolidated Afloat Networks and Enterprise Services.* San Diego, CA.

Ericksen, J., and L. Dyer. (2005). "Toward a Strategic Human Resource Management Model of High Reliability Organizational Performance," *International Journal of Human Resource Management.* Vol. 16, pp. 907–928.

Ewing, Philip. (2010) "Fewer Sailors Erode Readiness, Cut Ship Life," *Defense News.* June 21, p. 1.

Freeman, R. E. (1984). *Strategic Management: A Stakeholder Approach.* Englewood Cliffs, NJ: Prentice-Hall.

Friedman, Andrew L., and Samantha Miles. (2006). *Stakeholders: Theory and Practice.* Oxford University Press.

Hannan, M. T., and J. Freeman. (1984). "Structural Inertia and Organizational Change," *American Sociological Review.* Vol. 49, pp. 149–164.

Jansen, K. J. (2004). "From Persistence to Pursuit: A Longitudinal Examination of Momentum During the Early Stages of Strategic Change," *Organization Science.* Vol. 15, pp. 276–294.

Laplume, Andre O., Karan Sonpar, and Reginald A. Litz. (2008). "Stakeholder Theory: Reviewing a Theory That Moves Us," *Journal of Management.* Vol. 34, No. 6, pp. 1152–1189.

Legnick-Hall, M. L., C. A. Legnick-Hall, L. S. Andrade, and B. Drake. (2009). "Strategic Human Resource Management: The Evolution of the Field," *Human Resource Management Review.* Vol. 19, pp. 64–85.

Manacapilli, Thomas, Bart Bennett, Lionel A. Galway, and Joshua Weed. (2004). *Air Education and Training Command Cost and Capacity System: Implications for Organizational and Data Flow Changes.* Santa Monica, CA: RAND Corporation. MR-1797-AF. As of September 21, 2011:
http://www.rand.org/pubs/monograph_reports/MR1797.html

Miles, Raymond E., and Charles C. Snow. (1978). *Organizational Strategy, Structure, and Process.* New York: McGraw-Hill.

Mitchell, Ronald K., Bradley R. Agle, and Donna J. Wood. (1997). "Toward a Theory of Stakeholder Identification and Salience: Defining the Principle of Who and What Really Counts," *Academy of Management Review.* Vol. 22, No. 4, pp. 853–886.

Ostroff, C., A. J. Kinicki, and M. M. Tamkins. (2003). "Organizational Culture and Climate," in W. C. Borman, D. R. Ilgen, and R. J. Klimoski, eds., *Handbook of Psychology,* Volume 12: *Industrial and Organizational Psychology.* New York: Wiley. pp. 565–593.

Riposo, Jessie, Irv Blickstein, John A. Friel, Kindle Fell. (2009). *Navy Enterprises: Evaluating Their Role in Planning, Programming, Bdugeting, and Execution (PPBE).* Santa Monica, CA: RAND Corporation. MG-794-NAVY. As of September 1, 2011:
http://www.rand.org/pubs/monographs/MG794.html

Robertson, P. J., and S. J. Seneviratne. (1995). "Outcomes of Planned Organizational Change in the Public Sector: A Meta-Analytic Comparison to the Private Sector," *Public Administration Review*. Vol. 55, pp. 547–558.

Thie, Harry J., Margaret C. Harrell, Aine Seitz McCarthy, and Joseph Jenkins. (2009). *Consolidated Afloat Networks and Enterprise Services (CANES): Manpower, Personnel, and Training Implications*. Santa Monica, CA: RAND Corporation. MG-896-NAVY. As of September 1, 2011: http://www.rand.org/pubs/monographs/MG896.html

U.S. Air Force. (2008). Air Force Instruction 36-2640, *Executing Total Force Development*. December.

U.S. Air Force. (2009). *Career Field Education and Training Plan*. CFETP 3D0X2. August.